75 YEARS
OF SERVICE

**TO THE MEMBERS OF
MCLEOD COOPERATIVE POWER ASSOCIATION**

1935 TO 2010

75 YEARS OF SERVICE

TO THE MEMBERS OF
MCLEOD COOPERATIVE POWER ASSOCIATION

1935 TO 2010

BY SUE PAWELK

THE
DONNING COMPANY
PUBLISHERS

Copyright ©2010 by McLeod Cooperative Power Association
1231 Ford Avenue
Glencoe, MN 55336

All rights reserved, including the right to reproduce this work in any form whatsoever without permission in writing from the publisher, except for brief passages in connection with a review. For information, please write:

The Donning Company Publishers
184 Business Park Drive, Suite 206
Virginia Beach, VA 23462

Steve Mull, General Manager
Barbara B. Buchanan, Office Manager
Richard A. Horwege, Senior Editor
Jennifer Penaflor, Graphic Designer
Derek Eley, Imaging Artist
Tonya Hannink, Marketing Specialist
Pamela Engelhard, Marketing Advisor

John Richardson, Project Director

Library of Congress Cataloging-in-Publication Data

Data available from the Library of Congress

Printed in the USA at Walsworth Publishing Company

TABLE OF CONTENTS

PREFACE 6

ACKNOWLEDGMENTS 6

INTRODUCTION: FIRESIDE CHATS 7

CHAPTER 1	Life before REA	8
CHAPTER 2	Getting the Cooperative Started, 1935 to 1937	11
CHAPTER 3	Early Years Providing Service, 1937 to 1939	17
CHAPTER 4	1940s	23
CHAPTER 5	1950s	36
CHAPTER 6	1960s	44
CHAPTER 7	1970s	52
CHAPTER 8	1980s	58
CHAPTER 9	1990s	67
CHAPTER 10	2000 to 2009	77
CHAPTER 11	General Managers and Board of Directors	85
CHAPTER 12	Current Employees and Board of Directors	93

ABOUT THE AUTHOR 96

PREFACE

It has been seventy-five years since a group of members organized in McLeod County to bring electricity to their rural Minnesota farms. Their efforts in 1935 formed McLeod Cooperative Power Association, a rural electric cooperative, committed to the task of bringing electric power to the rural homes, farms, and businesses of its owner-members.

The Cooperative has grown to be the provider of electricity for six thousand locations in McLeod, Renville, Sibley, Carver, and other surrounding counties, and the provider of other services not readily available to the rural customer. In honor of the seventy-fifth anniversary year of McLeod Cooperative Power Association working to improve the life of its rural members, we have compiled this review of the Cooperative's rich history. It is a gift to each of our members who joined us to celebrate this milestone in 2010.

ACKNOWLEDGMENTS

Credit must be given to the late Frank Bargen, former publisher of the *Hutchinson Leader*, for compiling the first twenty-five years of McLeod Cooperative Power history in 1960; and to Linda Johnson, former Glencoe resident and writer, who edited the fiftieth anniversary booklet of the Cooperative in 1985. Thank you to Katie Ide for photographic contributions and to Pat Gavin for helping prepare copy for this seventy-fifth anniversary book.

INTRODUCTION

FIRESIDE CHATS

Tom Mix, *Gunsmoke*, and Franklin Delano Roosevelt's Fireside Chats. . . . Decades ago, they all came into the living room over that magic RCA Victor. Everyone "watched" the radio—but not always in rural America.

Rural electrification changed that.

Rural electrification brought the power, all right, but it also brought a new era in communication—first the radio, and now the wonders of satellite communications and cable television.

FDR, who signed the Rural Electrification Administration (REA) program into being in 1935, probably didn't realize what an audience he helped to create for his Fireside Chats—or, for that matter, what he had done to improve the quality of life in rural America. This is a commitment consumer-owned rural electric systems like ours continue to pursue.

CHAPTER 1
LIFE BEFORE REA

Before electric power came to a farm, adequate lighting was often not available. People were discouraged from reading because of poor light and they did not have radio. Farm people did not have access to information like folks in town. Even schoolchildren found homework difficult, as they had to compete for the light of the lantern with their siblings. The dangerous job of cleaning that kerosene lantern weekly had to be done or there was no light at all.

With no refrigeration, food spoiled quickly and diets were limited. Water had to be carried in buckets or if you had a hand pump on the back porch you were considered fortunate. Coal or wood stoves used for cooking and heating had to be refueled hourly and cleaned several times per week. Washing clothes was usually done outside in a kettle over an open fire. Water again had to be carried from a well or stream. Watering livestock was all done with buckets or pumped by hand into a trough.

Cows were milked by hand. Fresh milk had to be speedily delivered to the creamery before it would spoil since milk cooling was not available. Milking equipment had to be disinfected in hot water, probably over that outside fire. Farm families spent a lot of time chopping wood for stoves. Life was labor intensive and exhausting. Little time existed for socializing or entertainment. Before electricity, farm chores were all consuming.

Farming work was grueling physical labor before mechanization and electrification came to rural America.

Opposite page: For close to a hundred years the cast-iron cook stove was a fixture in American farm kitchens. In the 1930s, rural electrification offered rural families the promise of the convenience and cleanliness of the all electric kitchen. Gone would be the drudgery of chopping and transporting wood, the feeding of fuel and cleaning out of the ashes, the stove-top heating of water for dishes and bathing, and sweating over a hot stove in summer. Courtesy of Schwenkfelder Library and Heritage Center, Pennsburg, Pennsylvania.

Farmers shoveled grain by hand, ground feed by hand, and milked their cows by hand before electricity came to the farm.

In the early 1930s electrification of farmsteads seemed out of reach to farmers in this area. Only farmers living near a power line, such as alongside a major road, could get service. For most, the cost of constructing lines was prohibitive. Even if a farmer paid the $2,000 to $3,000 necessary to build the lines, utility companies assumed title to them. On top of that, rural residents still had to pay higher rates than their city neighbors. It was this situation that a few determined individuals sought to change as they looked for a way to bring power to the countryside.

The Rural Electric Administration (REA), created in 1935 by the executive order of President Franklin D. Roosevelt, was the vehicle they needed. Authorized by Congress in 1936 and reorganized under the Department of Agriculture in 1939, the REA instituted a program of long-term loans to state and local governments, non-profit organizations, and farmers' cooperatives.

McLeod Cooperative Power, the twenty-fifth association in the United States to apply for an REA loan, was organized in October 1935. But groundwork had begun months earlier and member-solicitation continued long after.

Photo courtesy of the Rural Electrification Administration

CHAPTER 2
GETTING THE COOPERATIVE STARTED 1935 TO 1937

According to early records, the first meeting to discuss rural electrification in McLeod County was at the Silver Lake Hall, on August 19, 1935. However, only a few farmers were on hand to hear W. R. Brabec, representative of the Emergency Relief Administration, unfold the REA plan. At a second meeting in Hutchinson the following night, there was again a disappointing turnout. Chester Graupmann, one of the early advocates, remembers planners were discouraged but not defeated.

Chester Graupmann

"They decided to have meetings in every school district. They planned to have a drive selling stock for $2 a share; three men in each township were supposed to do the soliciting." In Penn Township, Walter Radke, who later became a Board member, and his father, Adolph, worked on the sign-up. John Lipke put long hours in the campaign for members in Round Grove Township. Chester Graupmann set out with his father, W. W. Graupmann, and Art Schuette, to canvass Helen Township. Elvin Jensen and Lars Leifson led the quest for members in Bergen Township.

Volunteers were also at work in other districts. "We didn't miss a place. If they turned us down, we'd make a note of it and try there again." Graupmann said. One by one the signatures were collected. Graupmann noted with pride that organizers in Helen Township were among the first to complete a 98 percent sign-up.

A third meeting in the Courthouse at Glencoe in late September was packed. "Interest was keen," R. A. Fischer, the Co-op's first manager later wrote. Some, however, were dubious about a 100 percent loan and wanted time to investigate the details. Therefore, the Association wasn't officially incorporated until October 3, 1935.

Lars Leifson made the motion that a cooperative electrical association be formed. Following that he was promptly nominated and elected as temporary chairman. Incorporators included Lars Leifson, Glencoe; Walter C. Jungclaus, Glencoe; John Lipke, Stewart; Herman Graupmann, Biscay; and Charles Arlt, Glencoe. Capital stock was set at $4,000 and as originally planned, divided into two thousand shares at $2 each. Business was to begin when 20 percent of the authorized capital stock was subscribed and paid. They adopted Articles of Incorporation and By-laws, however, they were crude and later revised.

We were in business. Being among the first to form an electric cooperative, there was much to do; there was no pattern to follow. No one knew exactly what to do. We groped our way. We, however, grew from day to day—all we really had was a lot of determination, and believe you me, that is just what those early pioneers of this organization had.

All the secretarial work was done and directors meetings were held in the county agent's office. There was no money. The hat was passed among those assembled to which all contributed for the purpose of buying postage stamps. Meetings were held throughout the entire area; as many as four a day, to tell the people how they could obtain the advantages of electricity. We met in town and fire halls, schools, church basements, and homes. We had Service Agreements prepared, got as many to sign as we could and collected $2.00 memberships. Fortunately there were a lot of sincere people who were willing to work and pursue this cause.

The motion carried that the present Board of Directors be comprised of one representative from each township. Directors elected were: Acoma—Eric Tews; Hutchinson—B. F. Turman; Hale—Joseph Kadlec; Winsted—

Several members of the Board of Directors in 1940. They are, left to right: Carl Stender, Jake Hansen, Charles Arlt, Lars Leifson, Ben Peik, Herman Graupmann and Walter Jungclaus, with Elsa Pagelkopf, office secretary.

Theodore Ochu; Lynn—Virgil Jorgenson; Hassan Valley—A. W. Ohland; Rich Valley—Herman Graupmann; Bergen—Lars Leifson; Collins—C. A. Moore; Sumter—B. C. Peik; Glencoe—Walter C. Jungclaus; Helen—Chester Graupmann; Round Grove—John C. Lipke; Penn—John Schultz; and New Auburn—Charles Arlt.

The membership drive gained momentum and volunteers stepped up their efforts, seeking support for the fledgling cooperative. They knocked on the doors of every farm, hoping to persuade their neighbors that electricity was not only possible—it was necessary!

Elvin Jensen, who operated the Bergen Creamery, was also at work on the Cooperative's behalf. He frequently heard the farmer's concerns of a government take-over should the cooperative fail. Generally, he'd explain that farms would not be mortgaged to the government; officers and directors would not be responsible, as individuals, for repayment of debt. The electric plant, itself, and its equipment, would be collateral for the loan. Usually, those arguments alleviated concerns. One day, Jensen, said, he'd tried everything to allay a potential member's concern. In desperation Jensen finally said, "You know, if the others lose their farms for a $2 share of stock, you're going to lose yours, too. The government isn't going to let you sit here, just because you don't have this stock." The farmer signed.

Still, Jensen said, he sympathized with the farmer's reluctance to fork over the $2 fee. The price they got for their products was so small. Milk checks for a whole month, ran about $6. "To make matters worse," Jensen said, "at that time, we had a terrible drought. Farmers had to buy a lot of feed to keep their livestock. The price of feed was high. People had learned to do with very little."

In Renville County, Lynn Wulkan experienced similar difficulties. "We'd had a meeting, people would seem enthusiastic." Days later that enthusiasm would wane. "I was gone so much talking people into REA, that my dog barked at me when I got home each night. He didn't know me anymore," Wulkan said. Organizers tried to sign three farms to the mile, but Wulkan said, "We got as many as we could." Besides the $2 membership fee,

Lynn Wulkan

Ed Boyle

Arthur Ohland

farmers had to pay for the cost of wiring their buildings and contribute to the cost of line construction (an amount later refunded).

"Our biggest problem was always to get the required number of users," Wulkan said. "We had to get out and talk it up and really promote." When as many members as possible were signed-up in an area, the Board had to determine where to locate the line. MCPA initially contracted with Pillsbury Engineering Company of Minneapolis to make recommendations.

Ed Boyle (Uncle Ed to us), the jovial bachelor from Camden Township, was proud of the fact that he was the first person in Carver County to take out a membership in the Association. And he was largely responsible for extending our project into that county.

He had been reading about the beginning of rural electrification in the farm papers, and when he heard that this Association had been formed back in October 1935, he brought the matter up at a Town Board meeting. He and William Schmidt called on Lars Leifson and asked that the western part of Carver County be taken in. Next followed a meeting in District 50, which Leifson and W. R. Brabec attended and thirteen farmers promptly signed up, with Ed's name heading the list. The others were William A. Schmidt, Charles Schwartz, Joe Windmiller, Oscar Schrupp, August Maschke, Harvey Grimm, George Smith, August Dietel, and Ernest Ortlip all of Young America; and Ferdinand Heuer, Norwood. Mr. Boyle then attended the November directors' meeting, and he was appointed a director and asked to solicit members in Carver County.

When a member first came on, we had to arrange where to put the line and we had to get an easement. As in the sign-up drive, farmers had to be sold on easements individually. REA did not authorize use of loan funds to purchase rights of way.

Arthur W. Ohland was one of the pioneers in the formation of this Association, and credit for much of the early work should go to him. As project superintendent and later field man, he was the man outside and it was his responsibility to stake out lines, get rights of way, and obtain easements. In fact, he told a reporter that he "probably had gathered more easements that anyone in the country."

On February 19, 1936, McLeod Cooperative Power Association (MCPA) learned that its first loan of $600,000 had been approved by the REA. Directors could then look for a wholesale source of power. Several meetings were scheduled with representatives of Northern States Power Company (NSP), but the two firms couldn't come to terms.

The City of Hutchinson, however, in the process of building a generating plant, agreed to supply power at a rate about half that of NSP's and even purchased an extra engine to insure power for MCPA lines.

Bids for construction of the first section of line were awarded July 27, 1936. Directors chose to use western red cedar poles, General Electric transformers, aluminum conductors, and chose E. S. Gaynor Construction Company of Sioux City, Iowa, to do the work. The first pole set in the McLeod system was erected October 8, 1936, in Hassan Valley Township, about one and one-half miles south of Hutchinson. The first line was energized May 29, 1937.

Two days later, however, directors met and voted to terminate business with the Pillsbury Engineering firm. Among a long list of reasons cited were the problems experienced in energizing the first section of line. "When the first block was energized," Board minutes read, "fuses on transformers blew-out in 90 percent of the cases because fuse sizes were not ample to carry the load. This will necessitate changing all fuses which will cost the Association in the vicinity of $85, not to mention time and labor involved, nor the opinions that customers form of the project."

KEEP THE GRINDING JOB AT HOME ON THE FARM

The above picture is the same as that which will appear on our 1944 calendar, serving as a constant reminder of one of the many jobs electricity can do on each farm at a great saving of time and expense. This picture is of a 1 H.P. Viking feed mill, all set up with an over-head bin and hopper ready to grind feed every day in the year. No hauling or waiting in line or fussing to start the tractor in cold weather.

Customers may have been surprised by the fuse failures, but they weren't turned-off. By now, they were believers in electricity. Each month delegations would appear at Board meetings asking for electrical service to their areas as soon as possible. The word was out—farmers realized the potential for productivity possible with electricity.

Lynn Wulkan, talked about changes electricity made on his farm: "Before, everything was done by hand," Wulkan said. "We milked by hand; we hauled, pitched, and spread manure by hand from a wagon in the field."

Wulkan explained he had been "small and sickly" growing up. "Naturally, I was anxious to get labor-saving equipment. One of the first things I got was a used milking machine." Next, Wulkan devised his own barn cleaner, combining two long chains and a gutter with cleats and mounting the device on a homemade driving unit. "We pioneered in the use of labor-saving equipment. We had demonstrations out at the farm for feed-grinding. Between fifty to one hundred people would come to see this equipment in use," Wulkan said.

Jensen witnessed the changes, too. "Electricity made a day and night difference on the farm," he said. "They used it for everything." The creamery Jensen managed was wired for lights in 1937, but electricity never replaced cold well water for cooling milk there.

THIS MONTH WE CELEBRATE OUR FIFTH ANNIVERSARY

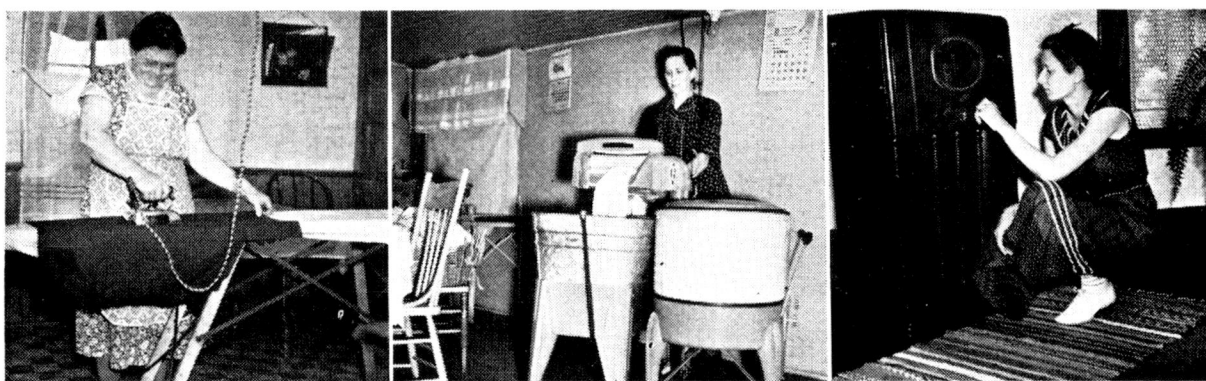

Upon studying our operating report this month one can readily determine that such constitutes our five-year record and in view of the fact that June, 1937, is our first monthly account, we, in reality, are celebrating our fifth anniversary. Most members being unfamiliar with our early history, a complete account of which would be interesting but too lengthy, we will reminisce only on a few limited portions and of the day we energized our line. Our association was organized October 3, 1935; the first bids to construct 112 miles of line were taken July 22, 1936, and the first pole erected October 8, 1936. All activities were centered on the day of energization. The engineer, the contractor, the construction foreman, and those of us who were actively engaged with the development of our organization were looking forward to this event with great anticipation. Questions naturally arose in our minds, such as, is the construction proper? Are all the connections made? When the energy is turned on, will the service continue uninterrupted? After all the work we had done, we felt it just couldn't be otherwise.

Our answer came on the eventful day of May 29, 1937, when the juice was first turned on. All was well! Electric lights for the first time appeared in many of our rural homes. Everyone was not only happy but many a housewife was thrilled to tears. She could visualize that times were changing; that the drudgery of farm life was being removed soon to become a dim memory of the past.

Bill Kircher of "The Farmer" of St. Paul was out that day taking pictures and interviewing members. Here is history in the making. Farmers owned and operated their own electric lines which was destined to become big business. The ladies began scurrying about connecting appliances to see if they really worked. Mrs. J. P. Karg of Hassan Valley township, appearing at the left in the above picture, was the first lady on our lines to do her ironing. Mrs. Herman Graupmann of Rich Valley township, in the center, was the first to do her washing and after having completed same, commented that she never before had done her washing with as much ease. The Walter Schultz farm in Hassan Valley township was also among the first to have their farmstead electrified. The daughter, Wilma, at the right, is shown turning in on a swing band, and thus our association first started operating. Let us ever be mindful of those days and cherish most highly that which we accomplished.

Wayne Bulau, past Board member, recalls the first appliance his folks purchased was an electric iron. "Ma didn't have to heat the irons on the stove anymore." The Bulau's pump jack and cream separator were soon motorized. "Also significant," Bulau said, "was the installation of a motor on the fanning mill for cleaning grain. It saved hours of standing there cranking."

Farmers weren't the only ones to benefit. The ripple effect spread to appliance dealers, who Jensen said, "had a regular heyday after REA came in." Association members purchased electric stoves, washing machines, refrigerators, radios, and smaller kitchen appliances, often before their homes were ever electrified, in anticipation of the power they had signed up to get.

CHAPTER 3
EARLY YEARS PROVIDING SERVICE 1937 to 1939

Wayne Bulau, longtime Board member and Bismark Township farmer, remembered when this pole was set and his farmstead energized on January 7, 1938. At that time the minimum charge for forty kWh was $3.25 per month (or slightly over 8 cents per kWh).

Progress continued in service distribution. Work began on a 362-mile section of line in spring 1937. Contractors for the $314,141 project were Martin Wunderlich Company and Acme Construction Company of St. Paul. At first, actual building of lines was tedious. "They hired men to set the poles by hand. It took a long time. It was a slow process to get going," Jensen said.

Wayne Bulau agreed. "Out here," he said, "all the work was done in the winter. Holes were dug by hand to put dynamite in. Then they blasted the frost away."

Poles were set in the Gibbon area in the fall of 1937. "January 7, 1938, we had lights in the house. The next day, we had lights in the barn," Bulau said. Many rural residents still remember the day they got electricity as easily as their own birthdays.

COWS WERE NOT SO SURE ABOUT ELECTRIFICATION

Mrs. Orville (Rosella) Lipke, wife of the former Board president, recalled that it was Christmas Eve 1937 at Orville's parents when Orville was young:

It was the first time they didn't get to church on Christmas Eve. They were milking cows and the electric lights came on for the first time, right in the middle of milking. The cows got so scared, they wouldn't let down their milk. It was a memorable night.

"WE KNEW ELECTRICITY WAS A GREAT THING. WE JUST HAD TO SHARE IT WITH OTHERS."

Delmer Schmidt, who lives north of Glencoe in Rich Valley Township relayed these stories:

In the late 1930s some guys would put up the poles. My brother Wilmer had gone to Dunwoody to be an electrician. He would run wires and solder the wires on the farms so they were ready to get electricity. He'd also rip out the upstairs floors of houses so they could get wire to the ceiling light fixtures. He'd run wire in the houses and install entry [fuse] boxes. Then the inspector would come and approve the wiring. He could always tell my brother's work because he used heavier wire than the rest of the electricians because he wanted to make sure people didn't have power quality issues.

There was one farm south of Glencoe in Sibley County, owned by Edwin "Eddie" Kelm. It was ready for inspection in 1940 except for the wire being soldered. My brother got sick and all of us in the family were quarantined for six weeks. We weren't supposed to leave the place. I never got sick or anything from my brother. But it was almost the Christmas season and this family could not get their lights turned on until the soldering was done and the inspection done. I had learned soldering and wiring from my dad who was a tinsmith. I snuck over to the Kelm farm and shimmied up the pole and soldered the wire with a gas torch. The inspector came, and the family had their lights turned on before the holidays.

Our family had electricity from NSP in a house we lived in prior to 1929. Then we moved to my parents' home place in Rich Valley Township, so we knew how electricity worked. When we got electricity brought in by REA to this farm, my family had an electric stove and refrigerator that my brother had already wired in. The electric stove was such an improvement for my mother every day but especially when she had to cook for the threshing crews. The old wood stove could never keep an even heat.

Most people knew nothing about electricity. My brother Wilmer built a tabletop demonstration to show people how a three-way switch would work. He'd show them how a yard light switch could be turned on or off from the house or from outside. This would help people decide how they wanted the wiring done on their farm.

APPRECIATING THOSE TWO WIRES

Buford Broderius remembered when REA crews came to the farm of his dad, William Broderius, Jr., in Melville Township in Renville County (where his brother Forrest Broderius still lives) during the summer of 1938. Harold Broderius had wired up the house so they could have electricity. Buford said:

REA crews cut down cottonwood trees by hand, with no chainsaws, along our grove. Dad and the hired hands dragged the trees into the grove to be cut up later. Crews dug holes and set poles by hand. After getting power, we really appreciated those two wires out there.

THE FESTIVE SEASON
By General Manager Ben Fischer

Thanksgiving is over, and it will soon be Christmas. Christmas to all of us has varied meanings, with the exception that it has a religious aspect. I would like to go back to the fall of 1939. We had let a contract to construct 117 miles of line from which we were to serve 397 members. Much had been accomplished during that period, and many of our people had wired their buildings, all in readiness to receive the advantages of electricity in their homes for the first time. Everyone was anxious to obtain electricity and particularly before Christmas. We just couldn't let these folks down, so the day before Christmas, a Sunday, we set forth to energize the farms that were ready—some eighty of them.

Hap Clayton and Fred Tillman worked as a team, took half of the number; Slim Meyer and myself (your manager) took the remainder. We started at the break of dawn, working primarily in Penn and New Auburn Townships. Many of the members during the morning were at church. We placed the necessary fuses and then energized the farmstead, checking as best we could to see that everything was all right. If possible we let the yard light burn, so that the member upon returning knew he had electricity.

We didn't stop for lunch, there just wasn't time. Will never forget, don't know the place, but somewhere in Penn Township we drove onto the farm yard about 3:00 p.m. There were a number of cars about the house. No one noticed our approach. We inserted the necessary fuses and energized the distribution system. When we were about to leave, a screaming mass of people came running out of the house, as though the very devil was after them. They had purchased a radio that was plugged in and turned on at full volume. Naturally, the radio program came blaring at them. They just didn't know what had happened. They were relieved and very happy when they saw us and realized what had taken place.

The last place we visited was in New Auburn Township. Darkness was upon us. We again drove up to the yard pole. I let Slim do the work and entered the barn where I saw a faint light. There I saw the husband and wife milking by hand with the aid of a dim kerosene lantern. I introduced myself and asked, "Your buildings are wired for electricity, aren't they?" They answered, "Yes." Then I said, "Why don't you use it?" Their reply was. "We don't have electricity." I said, "You soon will have." There was misbelief and silence. The barn in a few minutes was lit up and another family had lights on Christmas Eve.

We came home tired and completely bushed, but happy. We felt we did a good deed in bringing the first realization of electricity into these many homes and a great Christmas present it was.

Upon arriving home, we barely had time to take a bath, change clothes, then join our families in church and unite with them in Christmas services.

Fischer's article was published in *Our Line*, the Cooperative's newsletter.

A dedication picnic and program was held Sunday, August 28, 1938. About four thousand people gathered in Hutchinson's south park to celebrate the progress made by McLeod Cooperative Power Association. The picnic lunch was followed by speeches and music. Entertainment included the Hook-em Cow Quartet, the Hutchinson Band, Rudy Witthus of Buffalo Lake, the Winsted-Lester Prairie Trio, the Elling Brothers of Hamburg, and Marion and Vera Schmitz and Florence Buhr of Stewart. Renville county agent Frank Svoboda promoted a novelty boxing match to complete the program.

This nameplate was presented to Lars Leifson during the picnic and program at Hutchinson in 1938. It was attached to the first pole set, on the south half of the southwest quarter of Section 9, Hassen Valley Township, located a mile and a half south of Hutchinson. When the original pole was replaced, the plaque was brought in for safekeeping.

R.A. Fischer, then county agent, showed Walter Schultz of Hassen Valley how to read the meter.

Glenn Buck

To design the third block of line, and several projects that followed, MCPA hired Glenn Buck as engineer. An astute line engineer, Buck concentrated on efficient construction. "I never knew him to make a mistake," Wulkan noted. Like other REA engineers across the land, Buck was remarkably successful in simplifying construction and reducing costs. "The cooperative owes a lot to Buck's keen concept of economical line engineering," Harry Thiesfeld, MCPA manager from 1966 to 1977, remarked.

On February 23, 1939, at the fourth annual meeting, 410 stockholders were present. They changed the by-laws to increase the Board of Directors from nine to thirteen directors. They also consolidated the offices of secretary and treasurer of the Board of Directors.

In 1939, directors decided a full-time manager was needed. R. A. (Ben) Fischer, a vital member of the organizational team, was hired for the job. Offices were established in the Glencoe Community Building next to the Courthouse on Eleventh Street. Rapid growth and early loan pay back became part of the directors' philosophy. That fall, the Board of Directors stated its determination to expand in a resolution:

We the members of McLeod Cooperative Power Association, determined to increase the service of our cooperative to the farm families of our project area and to promote early self-sufficiency through maximum membership, hereby resolved to undertake a vigorous campaign to secure maximum additional member connections during the period starting December 8, 1939, and ending January 8, 1940.

CHAPTER 4
1940s

In November of 1940, the Cooperative launched a monthly newsletter to keep its members informed of progress of the Association. The first issue had a contest, asking members to choose a name for their publication. Sixty-five persons responded with many good suggestions for a name. The name *Our Line* suggested by Mrs. James Thelen of Hector, was chosen as the best name. Other names that made honorable mention were "The Ruralite," "Electric Flashes," "Illuminator," and "Current News." Our Cooperative has published a newsletter to its members every month since.

In 1940 the Cooperative also began offering cooking schools, to demonstrate to members how they could cook an entire farm meal with an electric oven, range top, roaster, and use of small kitchen appliances. Wives were urged to bring their husbands and members were encouraged to invite their neighbors who did not yet have electricity. Over one hundred women attended this first class in Buffalo Lake.

Many similar gatherings were held in member homes, where twelve families (husbands and wives) would meet at a member home to enjoy an electrically prepared meal. The ladies would gather at 4:30 in the afternoon and assist the REA instructor in the preparation of the meal. The husbands would come at 7:00 p.m. when supper was ready to eat. It gave the ladies a chance to practice cooking with the electric appliances.

Minnesota farmers who had already received power by 1941, were quick in putting it to work, judging by reports to REA. Dairying is a major farm activity, and almost a third of REA cooperative members have installed electric milkers. Forty percent of them have general-utility motors that can do dozens of jobs around the farm. Twenty percent are increasing winter egg production by poultry lighting, and there are many new feed grinders, stock tank heaters, water pumps, chick brooders, and other devices in use.

Eighty percent of housewives on Minnesota REA lines were doing the family wash electrically in 1941, and 87 percent had electric irons. In 28 percent of the farm kitchens electric refrigerators were providing safe and economical food storage. Almost 90 percent now had modern radios to bring them all the information and entertainment available on the air to those who lived in cities.

Most months in the early 1940s, *Our Line* reported who the largest kilowatt users were the previous month. This included farms, homes, and commercial users. Creameries were often located in rural areas and received electricity from REA. Even if they did not cool the milk electrically, they were sometimes large users such as Round Grove Creamery, Heatwole Creamery, and Lakeside Creamery. Businesses like Lakeside Store, Charles Borchardt (Metal Works) of New Germany, and Eddsville Store were frequently on the list.

The Cooperative's largest energy user during the early 1940s was the Platwood Club, a restaurant, nightclub, and bowling alley located along Highway 212 midway between Plato and Norwood (thus the name Platwood). The Platwood Club used between 3,000 to 5,000 kWh per month. The Platwood was built by Frank (Shorty) Dvorak in 1931. In 1942, he added four new bowling lanes. The place burned to the ground in February 1945, after a furnace kicking on supposedly ignited fumes from the bowling lanes being reconditioned. The *Waconia Patriot* reported the fire as a loss of $35,000.

With World War II, electric cooperatives throughout the nation experienced construction slowdowns as materials became scarce. But locally, the Cooperative trudged ahead. Art Sprengeler, former director, says Fischer's resourcefulness made the difference, "Ben was a good operator; he was foresighted. We moved ahead of other co-ops during the wartime. Materials were hard to come by, but he always had some on hand." In addition, McLeod Cooperative Power Association was in a solid position, having many miles of line already completed.

Lobbying efforts of Ancher Nelson, Fischer, and Lars Leifson in Washington, D.C., during January 1941, produced a $126,000 allotment for another 143 miles of line. When the bid came in well-below cost estimates,

30 more miles were included in the project. By the end of 1941 twenty-six hundred members were served on MCPA lines.

In March 1941, due to the needs of the National Defense Program, the Board voted to substitute copper conductors for aluminum; by November, it was a shortage of copper that caused construction delays.

In May 1941, faced with possible delays of construction materials due to the National Defense Program, a stock of hardware, transformers, conductors, etc., had to be carried in larger than normal proportions and especially so since it is necessary to change from aluminum to copper, which required duplication of materials. To insure warehouse facilities for such supplies and storage for trucks and other equipment, the Co-op saw fit to purchase the garage building known as the Home Garage at Glencoe. The savings in quantity purchasing of materials more than paid for the building.

A tiny baby girl was born in October 1940 to Mr. and Mrs. Donald Tieden, weighing just one and three-quarters of a pound. She was born at the home of her grandparents, the Ed Laffen's, of Melville Township, six miles west of Hector. This youngster caused Dr. Ralph Erickson, attending physician, great concern that proper conditions needed to be provided to preserve her life. Fortunately, the grandparents had a small battery set for light from which light bulbs were used in an improvised incubator to provide the necessary temperature. But it could not provide ample power or the proper humidity. The State Department of Health located an incubator for the family to use but they could not use it because electricity was not available in the home. The baby could not be moved to the hospital for fear she could not stand the change in temperature.

The doctor called the Cooperative Manager R. A. Fischer, asking if the Co-op could build a power line to the home. Renville County Welfare Board agreed that if the Co-op built a temporary line to the farm and furnished the material, that the County would pay for the labor to build the line and later remove it. Once the poles were transported, the line was staked, permission granted from the landlord and local electrician Harry Olson was busy wiring the house, the Co-op's eight linemen got busy building the line at noon. The entire nine-tenths of a mile was built (not according to specifications but serviceable) and energy supplied to the incubator at 5:45 in the afternoon.

Three months later, Our Line *reported that Virginia Mae Tieden was up to five pounds and would soon leave the incubator. In February 1941, the real-life drama of her earliest days was reenacted at McLeod Co-op Power's annual meeting by the Star Journal's radio drama group. WCCO radio recorded the presentation, hoping broadcasts of how electricity saved Virginia's life might help other premature babies. Virginia was interviewed in 1985 for the Cooperative's fiftieth anniversary issue, when she was a married mother of four.*

McLeod Cooperative Power Association received a Certificate of Merit from the secretary of the Department of Agriculture and REA administrator in September 1941. Only twenty-eight such certificates were issued nationwide and only three to Minnesota cooperatives. The award was made for achieving better than average success for the fiscal year 1939–40. The award was based on density of members per mile, consumption of electricity per member, prompt and complete principal and interest payments and no fatal accidents those years.

The annual meeting of 1942 boasted record attendance of fourteen hundred persons. Energy consumption was up 20 percent and revenue increased $16,000. Back then the Co-op hired a few ladies to prepare the meal for the annual meeting. It was interesting to note that to feed such a crowd it took twenty-three pounds of coffee, one dozen eggs, fifteen pounds of sugar, thirty-five pounds of onions, one hundred pounds of macaroni, five cases of tomatoes, forty-five pounds of hamburger, seventy loaves of bread, twenty-seven pounds of butter, sixty pounds of cookies, nine quarts of cream, three cases of pickles and condiments. Total cost of ingredients was $117.96 plus $15.95 for the ladies labor to prepare the meal.

Throughout the war years, REA promoted the causes of electric cooperatives, pointing to the need for electrically powered farm machines to replace manpower usurped by armed forces. Also stressed was the ability of electricity to improve farm productivity, providing more food for defense.

A huge diesel generator was installed at the Hutchinson Municipal Plant that served the Cooperative with power. In 1936 the Municipal Plant was buying three 450-horsepower generators and they increased that order to three 625-horsepower generators to accommodate McLeod Cooperative Power being their new wholesale power customer. On May 22, 1937, they started supplying MCPA with energy. The first day's load on our lines was 60 kWh and the peak was about 5 kW. It was not until 1938 that the peak exceeded the 500 kW limit in the contract, but long before this it was apparent that the load would grow rapidly. In May of 1938, the plant purchased a 1,500-horsepower engine and generator unit. It was the largest that could fit into the original building. By December 1940 the Co-op's peak had reached 960 kW demand. Early in 1940 the Cooperative's officers negotiated a ten-year extension of our contract with the Light and Power Commission.

Plans were immediately started by the City of Hutchinson for an addition to the plant sufficient to house three large units, and purchased the first of these. Such is the largest diesel engine in this part of the country, having a capacity of 3,060 horsepower. It is forty feet long and more than thirteen feet tall. At the rated load of 2,140 kW, the engine will consume about three barrels of fuel oil per hour and one gallon of lubricating oil. As of January 1942, the output to Co-op lines had frequently exceeded 10,000 kWh in a single day with peaks in excess of 1,000 kW.

At the request of the REA Defense Committee, armed guards were placed in 1942 about the Hutchinson Municipal Generating Plant and at our substation. The guards were on duty twenty-four hours a day and are instructed to allow no one on the premises unless they have official business there. We were at war and these precautions were taken throughout the entire United States.

In February of 1942, the Co-op's office located in the Community Building in Glencoe was enlarged and had more storage space added. A private office was made available for the manager. Lighting and heating were also improved. We were indebted to the Glencoe City Council for making these alterations for us.

The Cooperative newsletter often included not only progress reports on construction and new members added but also reports on safety and energy use. This is an excerpt from a 1942 *Our Line* newsletter:

ONE CENT COSTS $23.19

The saying "A penny saved is a penny earned" generally is true but not when the penny is used behind a blown fuse. When our service man, Hap Clayton, made his rounds reading meters last month, he noticed one that was unusually high and stopped to inquire whether or not additional appliances had been added. Being informed that none had, he surmised trouble and upon looking around noticed that the snow between the house and the garage was melted and also that the grass was already turning green. A portion of the farmstead wiring was underground and in this instance the cable not being buried deep enough was broken by the frost causing a short, which blew the fuse. Instead of remedying the trouble, a penny was placed behind the blown fuse, which held, but also consumed an additional 1,488 kilowatt hours of electricity at a cost of $23.19. Some penny! Moral: Never install larger fuses and when one fails, first find and remedy the cause, and certainly never insert a penny, always keeping in mind that the fuse is the safety valve to your wiring system.

Many members in the 1940s visited the Cooperative's display at the McLeod County Fair, where they featured homemade appliances such as chick and hog brooders as well as the fruit and vegetable dehydrator. A lighted project map also attracted a lot of attention.

Chick brooders were sold by the Cooperative as kits members could assemble or members could order them assembled for a fee. They were a very popular item in the 1940s.

The chick brooders, each with a five-hundred-chick capacity, were sold by the co-op with instructions for members to build it themselves or for an additional $26 to $31 the Co-op would assemble it for you. Since the number of heating elements available was limited, members needed to place their orders early. One year the Co-op built forty homemade chick brooders and they sold out very quickly.

Beginning on July 1, 1943, farmers had to get permits from the County War Board before they could secure copper to wire their farm buildings. Farmers had to apply for their allotment and their allotment would be deducted from our Co-op's share if they were our members. For members of our Association, not more than 67 pounds of bare copper could be assigned to each farm, which in most cases would do a pretty fair wiring job. It was serviceable but not as good as we were used to. Members whose farms were already wired but who wished to extend wiring into an outbuilding also had to apply to the War Board for materials. Third-quarter materials were meager with a total of 1,200 pounds of copper assigned to residents of McLeod County, 1,200 pounds for Carver County, and 750 pounds each for Renville and Sibley Counties.

If a farmer's buildings were not too far apart, 67 pounds of bare copper would give 220 service into the barn, wire the hen house, pump, and part of the house. Wiring could be extended into the house only if it was used for more than lights. Appliances also had to be wired up.

The above is a picture of the *Minnesota REA* Bomber which was purchased by Minnesota REA members through the sale of war bonds, Minnesota being the first state to thus qualify. Members of the Association made purchases in the amount of $24,575 during the drive. This bomber is the type that was used by the U.S. Army to sink the first Axis U-boat and which also was used by Major General Doolittle in bombing Tokyo.

The National Rural Electric Cooperative Association urged all REA members to support their sons, brothers, sweethearts, or friends in the armed forces by helping our government buy a fleet of bombers to help with the war. Here in Minnesota it was planned to sell enough war bonds to buy a $300,000 Mitchell B-25 bomber that would be named *The Minnesota REA*. McLeod Co-op Power members made bond purchases in the amount of $24,575 during the sign-up drive.

Due to tire and gas rationing during the war years, the Cooperative had to go to self-billing on January 1, 1943. Members had to begin reading their own meters, recording the number of kWh used, referring to a chart to determine the cost, and then sending the reading along with their remittance to the office. Or members could pay it at their bank. It was designed to save gas and rubber and also reduce the expense of billing, statements, envelopes, and postage.

In April 1943, of the 2,717 consumers served during the month, 29 overread their meters, 18 made errors in subtraction and paid too much, 17 paid too little, and 17 paid the gross amount instead of the net rate. Fifty-one failed to send in their meter reading or remittance. These would come straggling in during the next month. Back then the Co-op newsletter chastised those who did not get it right, stating that, "all the errors are inexcusable and it does seem strange that 2,656 members can get their reading in on time while 51 weren't able to. They forfeited their discounts, which for the year 1942 amounted to $664.07, a sizable sum that would have paid for a lot of electricity."

In 1944, Congress passed the Department of Agriculture Organic Act, known as the Pace Act. The bill extended indefinitely the lending authority of REA, which would have expired in 1946. In addition, the act changed the rate of interest on outstanding and future loans to a flat 2 percent. Previously, rates had been based on the government's cost of money, fluctuating from 2.46 to 3 percent. Payback period was extended from twenty years to thirty-five years.

After the war, REA gained permanent status. Congress had indicated strongly that all of rural America was to be electrified. Increasing amounts of capital would be needed, not only to finance new construction, but also to meet increasing demands of consumers already on the lines. Farmers were using more electricity each year. On a national level, the number of U.S. farms electrified climbed steadily from 10 percent in 1935 to 80 percent in 1950, 90 percent in 1955, and over 90 percent in 1959.

You never knew when the Cooperative personnel would be jumping in to help our members. In 1944, one of our members Alfred Hoefer, living a few miles west of Glencoe had his barn catch on fire. It appeared for awhile that the entire structure would be destroyed for the Glencoe Fire Department was using only one engine and the water supply was about gone. When, lo and behold, neighbors appeared on the scene with milk cans and stock tanks filled with water, which were emptied via buckets into the engine's water supply tank. A mad rush was made for more water. Firefighting apparatus from Brownton and New Auburn were called to the scene and proceeded to pump water from nearby potholes until the blaze was under control. One end of the barn was badly damaged, but it was far from destroyed. Our servicemen, Hap and Lee, did their bit by climbing to the top of the round roof with the aid of their climbing hooks and manned the hose. Hoefers appreciated all the cooperation. The fire, by the way, was not caused by defective wiring or any part of the electrical equipment.

As early as 1945, MCPA consumption approached that of its supplier, the Hutchinson Municipal Plant. "Shortage of power was getting to be a problem," Art Sprengeler, who served on the Board of Directors from 1944 to 1953, observed. An appeal was made to reduce peak loads. Demand was so great during milking times that the power would be knocked out. A power-rationing system, in which farmers rotated milking times, solved the problem until some farmers did not adhere to their designated time and the line would go out again. In early 1947, members efforts to reduce the peak, kept the system's demand level at 2,740 kW, which otherwise would have exceeded the plant's 3,300 kW capacity. This situation was alleviated May 2, 1947, when the Winthrop Substation was energized, allowing MCPA to obtain power from a second source, Northern States Power.

At 6:00 p.m. on August 14, 1945, it was a memorable moment for those listening to the radio, as Bill Henry of the National Broadcasting System advised the nation that President Truman had announced the Japanese Imperial Government has accepted the terms of unconditional surrender to the Allies. The world was again at peace. What a wonderful sensation after years of bitter strife. The following day was declared a holiday. However, V-J Day had not been officially declared until Japan actually signed the peace documents.

In the mid-1940s the Cooperative began offering free welding schools. The Co-op would host one-day training a few times a year, for farmers to come to the REA warehouse for welding classes. Initially they taught how to use a welder. Later, in about 1947, when more farmers had welders, they focused on helping members with their welding problems. Members could bring in broken parts to get them welded during the demonstration, as time permitted. Classes were taught by representatives from welding machine companies.

Ervin Westphal reported on electrification of their farm place that was owned by his dad, Ferdinand Westphal, in the early 1940s:

My dad helped the Co-op put in the poles by hand because they wanted power in before the holidays. My mom had the toaster plugged in waiting for the power to come on. As soon as it came on, she was making toast and dad was shaving with his new Remington electric shaver. The next night, dad threw the old razor and soap away.

Some people tried to keep the light bill below $3.60. Some of our neighbors used 25 watt bulbs to keep usage low but dad made the electrician wire for all 50 to 60 watt bulbs to make sure we had plenty of light to see what we were doing. Our yard light was a blessing too.

Gloria Stark Pudewell remembers when power came to their farm:

Electricity came to our farm in Bismark Township in Sibley County, on December 24, 1941. The reason I know the exact date is that it was my father's birthday. I was just five years old. I do not remember us having lamps, but I do remember the lights coming on that night.

Lorraine Wetzlaff, daughter of Albert Pitzner, who resides one mile from her home farm in Dryden Township, recounted when they got power in 1942:

The line to our place was the last line built before World War II. Dad got the last refrigerator available out of Gaylord before the war. The fact that we had a refrigerator instead of a block of ice or putting the milk and butter in the well pit was pretty neat. I also remember us having a waffle iron.

The ceiling lights sure beat the kerosene lamp for safer and better light. It lit up the whole room. I was twelve years old at the time and it sure was nice because I didn't have to clean the kerosene lanterns anymore.

SCHOOL 3800th CONNECTED MEMBER

As is customary, we make a special announcement of every 100th member who has received service. We are very pleased to inform our readers that our 3800th user is Renville County School District No. 133, located in Section five, Brookfield township. No, this district or school board was not lax in having electricity sooner. They did want it, but there was a matter of cost as the school is located one-half mile from our existing line and as there are no farm places close by, their rates, therefore, became greater than for most other schools.

Miss Lippert and pupils of School District No. 133

The school board comprised of Roscoe Grams, chairman; Clarence McBride, clerk; and William Hackbarth, treasurer; decided that the pupils in this district should enjoy the benefits of modern lighting in their school and made application for same. The service was completed last December and electricity was first used for the Christmas program, in fact, our construction crew and the electrician, Hans Petterson, did some special hustling to have the job completed for the occasion.

This school, a one-room structure, is well illuminated with four fluorescent fixtures. The teacher, Miss Phyllisann Lippert, whose home is at Olivia, notes an increased ability on the part of the pupils to study, due to adequate lighting. Their posture has improved and they experience less headaches. Lights, during the winter months, are frequently used the entire day.

Eleven children are at present enrolled. They are: Diane M. Grams, Doris M. Hackbarth, Dorothy Zachow, Glen Dovenmuehle, Ronnie Grams, Margaret Hackbarth, LaVonne Varland, Donald Carte, Glendon Zachow, Une Dovenmuehle and Martha Hackbarth.

We hope you all enjoy electricity in your school and will use it to prepare hot lunches, too.

Electric Lights Make for Better Studying

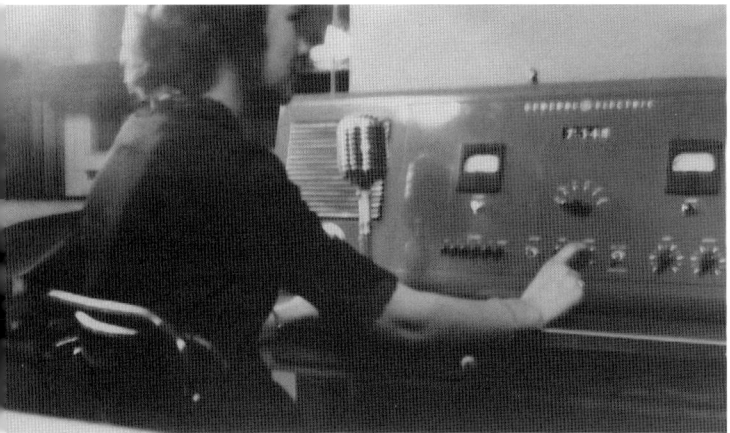

Above: Our Cooperative was the first REA and second organization of any kind to use a two-way radio system in Minnesota. It was October 2, 1946, when the Co-op first talked to the trucks in the field.

Top right: Linemen could talk twenty miles truck-to-truck. Sometimes messages had to be relayed via the office if trucks were farther apart. In 1956, a new Motorola private line radio system increased their range to fifty miles. Foreman Hap Clayton is shown with the new radio installed in his service truck.

Bottom right: Trucks were outfitted with the radio transmitting unit behind the cab.

In December of 1945, the Association crews accomplished a notable feat of setting 300 poles in nine days—an average of 33.3 poles per day. This in itself was no great accomplishment except that the poles had to each be set in one to three feet of frost, and the job was northwest of Hector and took extra travel time each day. To get the job done the Cooperative was able to hire a power digger from Kandiyohi Cooperative Electric Power Association of Willmar.

After using the digger, MCPA seriously considered purchasing one of its own, as it made construction go faster, eliminated a lot of hard work and was invaluable for moving poles on new road construction jobs. But upon investigating the cost of a new commercially built digger, our Board of Directors and manager readily gave up on the idea as too costly. They did not despair, however. They knew that the government had purchased similar equipment during the war and they made an inquiry about war surplus units. They found out that several were available. Board President Herman Graupmann and Foreman Hap Clayton flew to Washington, D.C., and drove home a new digger from Baltimore that had never been used. It was mounted on a one-and-a-half-ton Chevrolet four-wheel-drive truck that had only been driven 712 miles and the entire equipment was purchased at half price. MCPA line crew got used to using the truck quickly and then the staking crew had trouble keeping ahead of the pole setters.

In 1946, the innovation of a radio system improved the efficiency of service. Our Association was the first REA in Minnesota, and the second organization of any kind in this state, to use two-way radio for communication purposes. October 2, 1946, was the first day we talked from the office to the men in the trucks, and they in turn, between themselves and back to the office. The first installation was the General Electric FM system. The trucks ordinarily communicate between themselves at distances of up to twenty miles and in the event the communication is not clear, the message must be relayed from the central office.

In October 1956 a Motorola private line radio was installed, with the sets in the trucks being increased from 30- to 100-watt units. This kept out all outside interference and the range now was at least fifty miles instead of a poor twenty. We were pioneers in the field, being the first REA in the state to use this equipment. Since then radio had come into wide use among law enforcement officers and others.

The monthly cooperative newsletter in the 1940s carried safety warnings for members, just like today. In 1944, members were warned about the hazards of homemade electric fences, stating that many members merely connected their fence to a 110-volt circuit and use a 7.5-watt bulb as a resistor, thus conveying the full voltage through the fence. This condition was very hazardous to anyone coming in contact with it.

The danger of this type of fence was more forcefully brought to our attention through an incident related to us from a member in Collins Township who had a fence connected as above mentioned. Two of his boys, ages seven and ten years old, were playing within the vicinity of this fence. The older boy dared his brother to touch it. Upon doing so and grasping with one hand, he instantly became frozen to it and with the other hand grabbed a steel post nearby, forming a perfect ground. The child was helpless. Fortunately, the older boy had presence of mind to disconnect the current in the henhouse, thereby saving his brother's life. This was a close call and reminder that only fencers with intermittent contact should be used.

The Winthrop line was energized on May 2, 1947. Power was turned on to seven farms on a trial basis. On May 5 everything operated smoothly and farms in the southern area were fed from the new substation, including farms in Bismark, Transit, and Dryden Townships and a portion of Round Grove and Penn Townships. This greatly improved the quality of service for some members. A second new substation was energized at Bird Island later that year to serve Osceola, Melville, Preston Lake, Round Grove, and Penn Townships.

The Cooperative was often using its newsletter to set the record straight, be it regarding investor owned utilities circulating incorrect information about the Co-op or about how much McLeod Cooperative Power paid in taxes. They detailed for members in *Our Line* just how much the organization paid in taxes in 1948. They noted that taxes would be more in 1949 and 1950 after their current construction program. The total paid in 1948 was $4,374.88 broken down as follows:

Personal Property Tax	$ 1,501.50	State Corporation Tax	59.81
Real Estate on Glencoe Warehouse	156.82	Social Security Tax	382.74
Pole Yard at Glencoe	9.30	U.S. Excise Tax	155.82
Lots for Building Site in Glencoe	129.52	State Unemployment Tax	191.36
Hutchinson Substation	395.62		
Winthrop Substation	199.06		
Bird Island Substation	165.32		
Winthrop Transmission Line	441.79		
Bird Island Transmission Line	163.83		
Building Lot in Hector	12.39		
State Membership Tax	410.00		

In another issue of *Our Line* the editor shared another ditty to set the record straight:

Private utilities make a big fuss about REA being subsidized by the Government. We are told that the money borrowed from our Uncle Sam upon which we pay 3 percent interest costs him less than 1 percent, so it would appear that we are instead subsidizing the Government. Those big boys do like to get us confused.

In 1948, the Cooperative completed a system study of its project. The purpose was to make sure it had a distribution system that was adequate to carry the electrical load of the members. It included making sure substations were properly located and that conductor sizes were of large enough carrying capacity. Proper sizing and placement of circuit breakers, voltage regulators, and line fuses were also checked.

In making this study, township maps were prepared on which the farms of all the consumers were located, as well as those farms and residences yet without electricity. These maps gave Cooperative engineers a picture of what had been accomplished and what remained to be done.

Interestingly, the study disclosed that 91.5 percent of the farms, schools, churches, town halls, creameries, and county stores along our line, inclusive of the territory we serve, were now enjoying the benefits of electricity. Such a record is remarkable when we consider that eleven years prior to 1948, relatively few of the Co-ops 4,090 consumers enjoyed the privilege of highline service.

Prior to December 1941, contractors had been hired to build the Co-op's project. They built 963 miles of line to serve 2,226 members. From late 1941 to 1948 the Co-op supplied its own personnel and equipment for construction. They had built and rebuilt their own substations, transmission lines, and distribution lines, at a savings to the members. The busiest years for MCPA's own guys were 1943 to 1947 when they built between 118 and 256 services each year. After 1948 the demand for farm services slowed down. In 1948, the Cooperative tried to extend service to the remaining 367 nonelectrified locations in the service area, just as quickly as they would apply for service. Hale Township had the largest number of remaining nonelectrified farms and homes at 66.

During 1948 McLeod Cooperative Power took over the distribution lines of New Auburn, serving 110 accounts and nineteen miles of line and also 55 rural consumers that we obtained through the Eastern Minnesota Acquisition. It brought our total number of consumers to 4,033. The Co-op made many improvements to lines within the Village of New Auburn and extensions to farmyards and buildings.

Less than one year after acquiring New Auburn, it was sold to Northern States Power Company. On March 15, 1949, NSP purchased the city services, as the Co-op believed the Rural Electrification Act was designed to bring power to rural consumers, not small towns, thus the reason for the sale. The Co-op retained all rural accounts not in the Village that it got through the Eastern Minnesota Acquisition.

A new office and warehouse were also in the works. A building site at the corner of Ford Avenue and Highway 22 was purchased in 1947 and the Minneapolis architectural firm of Long and Thorsov designed a one-story "semi-modernistic" structure of concrete, steel, and brick, with a 126-by-30-foot office, 117-by-76-foot garage, and 40-by-60-foot metal warehouse.

Melvin ZumHofe, Glencoe contractor, submitted the low construction bid of $129,184 but additions to plans brought actual costs to $134,500. Construction began in 1949.

Sketch of new building designed by Long and Thorsov of Minneapolis. It was semi-modernistic design and was featured in the *Minneapolis Star Tribune* in 1949, along with other Cooperative headquarters being built in Minnesota.

The site of the Cooperative's current home at the corner of Highway 22 and Ford Avenue in Glencoe, when initial excavating work was being done to prepare for construction of the building in 1949. A massive black walnut tree was removed from the site during excavation and used to make the two board room tables still in use today.

CHAPTER 5
1950s

Quarters in the Community Building, which housed the offices until the move to the new building, had become so crowded and cramped that the Board of Directors and management had long deliberated about the necessity of providing suitable quarters. Directors first met at the new building in March 1950 to plan furnishings for the boardroom. Lynn Wulkan remembered that they decided to have two 5-by-8.5-foot solid tables crafted from a mammoth black walnut tree cleared from the building site. Those tables are still in use today, and can be seen in many of the Board photos.

Cooperative employees moved into the new building in 1950. About three thousand people toured the building during its Open House celebration in March of 1950.

An Open House officially celebrating the completion of McLeod Cooperative Power Association's new office building and warehouse was held on Wednesday, July 19, 1950, between the hours of 10 a.m. and 9 p.m. with some three thousand people taking a tour of the newly completed facilities. More than two dozen beautiful bouquets of flowers sent by well-wishers and friends added greatly to the occasion. Directors and employees and their wives served as hosts and receptionists on this red-letter occasion. Coffee and doughnuts were served to the many persons who came to inspect the new facilities.

Visitors were impressed with the roominess of the building, the exterior of which is deceiving as to its size. As guests entered, they registered in the lobby, and then viewed the director's room. They then went to the billing room where the posting machine was explained. Then to the cashier's and receptionist's desks, and at the latter, features of the two-way radio and the inter-com systems were reviewed.

They then inspected the manager's room, the fireproof vault, bookkeeper's room, general office, mailing room, engineer's room and then saw the furnace and air conditioning equipment in the basement, then back upstairs where they toured the rooms of the linemen, meter testing, small parts department and into the warehouse where the equipment and trucks are stored. Features of the warehouse include the overhead tracks, which lead from the loading dock and extend over the entire length of the building. Equipment included a two-ton electric hoist, which greatly facilitated the loading and unloading of heavy equipment. Of prime interest was the snowmobile, which impressed many visitors who saw it for the first time, and which proved to be indispensable during the heavy snows of the following winter.

The July 1950 Board of Directors meeting was held at the new headquarters. Directors found it pleasant to meet and transact their business. The chairs were comfortable, the acoustics and lighting good. Former directors were invited to attend as guests and sit in on this first meeting.

Less than a month after the new office opened, the display area in the lobby was already graced with an array of electrical merchandise from appliance merchants in the area. This space was made available to all

The Cooperative had a Bombadier snowmobile that was first put to use in spring of 1951. Sometimes crews use it to get to places when lines were down when roads were otherwise impassable. It was also put to use bringing doctors to patients in bad weather and once to bring in enough Board members to make a quorum for a meeting.

Board Members Meet in New Office

Former Board members joined the active members for their July 1950 meeting in the new office. On the picture, left to right, seated: Victor Hahn, Stewart; Theodore Dietel, Glencoe; George Lhotka, Silver Lake; Walter Jungclaus, Glencoe; Lynn Wulkan, Hector; Ed Boyle, Norwood; Arthur Sprengeler, Green Isle; Melvin Todd, Brownton; Alvin Fischer, Buffalo Lake. Standing: Hubert Smith, attorney for the Co-op; Jacob H. D. Hansen, Hutchinson; Harry Bulau, Gibbon; R. A. (Ben) Fischer, manager; John Lipke, Stewart; Arvid Anderson, Hector; Ben Peik, Brownton; Glen Buck, engineer; Lars Leifson, Glencoe; and Frank Haas, Buffalo Lake. Only director missing from the picture is Ancher Nelsen, Hutchinson. Photo by Duke Thiele of Glencoe.

dealers of electrical appliances in the Co-op's trade area. Each dealer was given space to display their wares for one month free of charge. It gave members the opportunity to see what new appliances area vendors had for sale when they were transacting their business at the Cooperative office.

Beginning in June of 1950 the Cooperative office was no longer open Saturday afternoons. Previous to this time the office had been open six days a week and crews worked a six-day week also because of heavy construction. Office people would only work Saturday mornings as an economy measure. Work for line crews is more routine now and they began working a five-day workweek, except for responding to outages and emergencies.

The fifteenth annual meeting of the Association, held at Glencoe High School Auditorium, was well attended. It was estimated that four thousand persons were in attendance at the meeting and electric show. The Co-op had provided a noonday lunch for twenty-seven hundred persons at the Glencoe Fire Hall with serving done under the supervision of the Glencoe Civic and Commerce Association.

By 1950, the Association ranked second in advance payments of principal ($300,000) among all 936 REA projects in the United States. The McLeod Cooperative Power Association was in good financial standing. They, in fact, could have repaid their loan to the government any time they should have chosen to do so. They instead believed it more advisable to have substantial reserves on hand with which to meet emergencies and enlarge their system without additional loans.

Local appliance dealers and farm equipment vendors displayed their wares at the Cooperative's annual electrical show in 1952.

Throughout the next decade, growth could be charted each year, as consumers took advantage of low rates to exploit the labor-saving possibilities of electricity. Revenues for 1951 reached the half-million mark and MCPA continued to expand.

Our Line reported that a car crashed into the new building. On January 4, 1951, the driver of a car who had indulged freely from the bottle, crashed into another vehicle at an intersection north of Glencoe. In his haste

CHAPTER 5 1950s

Mrs. Martin Schwarzrock and her daughter Donna were third-place winners in NRECA's national song writing contest. They were from Hector.

to get away from the scene of the accident, he drove at a high rate of speed, through the stop sign at the intersection directly northeast of our new building. Upon doing so, he above all things, hit a state highway patrol car, and sent it careening across the icy street straight toward our display window!

Patrolman Barnes, who was driving the state car, had presence of mind to crowd the wheels to the left. The left wheel hit the north steps of our building, and the car stopped with a terrific impact against the steel pillar of the north planter box. Barnes said, "That window looked awful big for awhile."

The patrol car was a virtual wreck and sold as junk. The building wasn't badly damaged. Several bricks in the flower box were broken and others chipped. No one, fortunately, was injured.

In 1952, a forty-by-forty-foot office/warehouse was built in Hector to service western-area consumers. Throughout much of the 1940s and 1950s the Cooperative had at least one, and sometimes two, servicemen stationed in Hector to respond to calls in the western portion of the system.

In 1953, the president of the United States selected one of the MCPA Board members to serve as rural electrification administrator in Washington. Ancher Nelson, served as a director of MCPA since 1936 and until this appointment to REA. He was Minnesota lieutenant governor at the time of his appointment.

The Cooperative participated in a songwriting contest in 1953 that was sponsored by the National Rural Electric Cooperative Association. We received two entries from MCPA members, which were both submitted to the national contest. One song, from the Martin Schwarzrock family of Hector, tied for third place in the national competition from six hundred entries. They won a prize of $25. Mrs. Schwarzrock wrote the lyrics of "One Happy Swede" to the tune of "Rueben and Rachel" one evening, just as a lark. She thought it was too silly for a mother of three children to enter the contest. Daughter Donna did not agree and told her mother that she would submit the song and collect any prizes that came their way. Their worthy entry went like this:

One Happy Swede

1. *Aye ban a farmer in Minnesota, ban here mostly all my life. Got myself some fancy contraptions, mostly to please my purty wife.*
2. *The R.E.A. she's sure ban helpful, made our job here on big fun. All we do is press the buttons, then, our work, by gosh, she's done.*
3. *Got some milkers, sure are tricky, to my cows I put them on. Turn around and pet the calves and then, by gol', my*

milkin's done.

4. *Do I stand and pump my water. I should hope to tell you not. Push the lever, chew my snoose and then, by gol', the water I got.*
5. *Does my missus carry corncobs, chop her wood and get all in, I should say not and why should she, with her 'lectric stove, by gin.*
6. *Now that woman doesn't nag me, for the water to go and get, all she does is turn the faucet, and the whole durn place gets wet.*
7. *When she turns the 'lectric bulbs on for the neighbors all to see, then I tell you this ol' farmstead shines just like a Xmas tree.*
8. *All I need to make me happy, is that thing they call TV; with the money Our Line saves me, I can get one soon, by gee.*
9. *Now this R.E.A. is something, with it you just can't go wrong. But I tell you, and I mean it, this ol' Swede helped things along.*

This Year's Float

The above float already has or will appear in the summer festivities featured at Hutchinson, New Auburn, Glencoe and Winsted. Seen on the float are Lois Jungclaus and Pearl Stender of Glencoe, and Diane Ortloff of Plato.

At the twentieth annual meeting in 1955, net margins were reported at $121,000 from $686,000 of total revenue. Total revenue two years later was $766,488; over 30 million kW of electricity were purchased, representing an 8.7 percent increase in consumption from the previous year. At the annual meeting in 1957 the first batch of capital credits were refunded.

A big sleet, rain, and snowstorm on March 14, 1957, caused many power outages in the Co-op's service area. Visibility was zero. The only way linemen could drive was with their heads out of the window watching the shoulder of the road. Often they did not know where they were. Their comment was "God bless you people who have your names on your mailboxes." The radio system also proved to be invaluable at reporting progress made, reporting damage and equipment needed, and saving a lot of time.

The MCPA office was no longer open Saturday mornings beginning in December of 1957. The office would however be open from 7:00 to 9:00 p.m. on Friday nights during the collection period, that is any Friday night that fell between the fifteenth and twenty-fifth of the month. The Co-op was following the pattern of local merchants whose stores were open Friday nights and remained closed all day Saturday.

Self-billing continued throughout the 1950s, 1960s, and to mid-1971. Members would be mailed a calendar with a pocket each January. Inside the pocket was a year's supply of meter cards individually stamped with the consumer's name and account number. Members also received a rate card to help them know how much to

Calendars, Calendars

Many hours were spent by the staff and extra help in preparing envelopes with the 1966 calendars and meter slips which have been mailed to each McLeod Co-op Power Ass'n. member. A total of 4228 calendars were mailed early in January in time for the meter slips to be received well before the January 15th meter reading date. If one person had done all the work of putting addresses on envelopes, stapling the meter slips together, filling the envelopes, tying the bundles for mailing and taking them to the post office, it would have required at least twenty working days. This annual undertaking grows each year as the cooperative membership increases.

Mrs. Muriel Von Berge, Marilyn Rannow and Nancy Nemitz form an assembly line to prepare the envelopes for mailing. (Continued on page 2)

CALENDARS, CALENDARS
(Continued from page 1)

Bundled envelopes are stacked awaiting delivery to the post office.

Max Hoxie and Les Schrupp load over 1000 pounds of envelopes containing calendars and meter slips.

charge when they figured their bill each month. Those few members that made errors by paying too much or too little, would receive a notice from the office employees at the Co-op instructing them how much to add to or deduct from their next bill to correct it.

During the 1950s members were also encouraged to install a control on their water heater. A cheaper off-peak rate of 1.43 cents per kWh was offered to participants for their separately metered water heater use.

Construction continued at a steady pace. Work began on twelve miles of 69,000-volt transmission line to serve the east half of the project area. The Bell Substation, 3.5 miles south of Hutchinson, was also started in 1958. Construction costs for both projects, totaling $195,000, were paid out of operating funds.

The mid-1950s meant a major boon to Minnesota cooperatives when "low cost hydro power from the Missouri River became available through the U.S. Bureau of Reclamation." Cooperative Power

Garrison Dam under construction in western North Dakota. It would generate hydro power from the Missouri River to benefit those cooperatives and utilities receiving its low-cost kilowatts.

Association (CPA), organized in 1956 as a continuation of the Power Development Committee, was comprised of eighteen distribution cooperatives uniting to boost their bargaining power. The wheeling contract negotiated by CPA, allowing cooperatives to use the transmission lines of private power companies to receive inexpensive power, was a landmark change.

> **WANT TO REDUCE YOUR ELECTRIC BILL?**
> **And We Quote:**
> Most folks waste about 15 per cent of the electricity they pay for. We want to be the first to tell you how to stop this waste:
> Keep your wife from setting the furnace thermostat too high. Train her to keep on the move and keep warm nature's way.
> Be sure your wiring is heavy enough to carry the power you need. Electricity wastes itself trying to crowd through small wires.
> Cook everything rare. A little step each day and you can soon eat it raw.
> Be sure that wiring does not come into contact with water pipes, eave spouts, trees or roofs of buildings.
> Marry girls off young. Let some other dope heat their irons.
> Give motors and equipment a periodic inspection. The squeaky wheel uses more kilowatts.
> Go to bed earlier.
> Place your water heater as near the sink, lavatory and bath tub as possible.
> Use small pipes and insulate well on long runs.
> Don't bathe your kids often. If they smell, send them outside to play.
> Stop all faucet leaks. You pay to pump and heat water. Why waste it?
> Ditch your TV (not a bad idea!)
> If a fuse blows, find the cause and correct it.
> Don't pay your electric bill and be disconnected. This will save everything.
> Pay our electric bill before the 25th of each month and save the penalty.
> Go to church frequently, while there, your TV, radio and other equipment will not be in use.
> Locate your refrigerator and freezer where they will not be in a direct blast of your heating system.
> Visit your friends and relatives as often as possible. Accept their invitations to stay for meals—even for baths if invitations can be wrangled. Their water is just as good as yours.

As a result, Bureau of Reclamation power coursed into southern Minnesota, bringing with it a tide of rate reductions. In 1958, *Our Line* newsletter reported "bureau power" meant savings of $4,833 for MCPA during the first month. Savings were passed on to consumers. In 1959, a 9 percent rate decrease saved members about $75,000.

Through the years line workers had responded to many outages and many times storms had damaged specific areas. *Our Line* seldom made a big hoopla out of any of the storms except to routinely report on what happened. In the fall of 1959, they reported it as "our worst storm." It apparently was a storm that raged for several days and caused undue hardship for many of the Co-op's members. More than three-fourths of the lines were out of service. The storm started as rain on Saturday, which gradually froze and changed to a wet snow Saturday night and Sunday. Power was out in some areas Sunday though Friday. Carver County, central McLeod County, and Sibley County were hit much worse than Renville County. Lines were heavily loaded with up to three inches of ice weighing three pounds to a foot. On a three-hundred-foot span with two wires that would be eighteen hundred pounds of load or more than three times the rated carrying capacity of the conductor. MCPA never lost a pole on the entire system, but there were breaks in lines, and three transformers burned out.

Transportation was the greatest problem. The roads were impassable and made patrolling lines nearly impossible. By Tuesday forenoon some roads were open and service crews could wade through snow up to their hips or get into farmsteads on horseback or by sleigh. Communication was completely out so crews would have to come back to the office to relay progress made.

Want to reduce your electric bill? The editor of *Our Line* in 1959 had some suggestions (see above). I doubt we would make some of these suggestions today.

CHAPTER 6
1960s

The first electrically heated home on the MCPA system was the home of Leon and Arlene Schuft and their daughter Kimberly. They lived five miles southwest of New Auburn. They installed glass-panel radiant-type heating units during 1960. Baseboard heaters were used in the kitchen and living room and had separate thermostats. Wall-mounted electric heaters were used in the bathroom, porch, and bedrooms. Following this first installation, electric heating rapidly caught on throughout the Co-op's service area.

The 1960 Board of Directors gathered in the board room with its room-width mural wall. Seated around the table are: Jacob H.D. Hansen, vice president; R.A. Fischer, manager; Lynn Wulkan, president; Ed Boyle, secretary-treasurer; Carl Ortloff; and Walter Radke. Standing: Glenn Buck, engineer; George Jungclaus; G. E. Birk; Victor Pulkrabek; Wayne Bulau; Hubert Smith, attorney; Arvid Anderson; Orivlle Lipke; Clarence Walter; and Theodore Dietel.

Above is our 25th anniversary float which was part of the Hutchinson JC Water Carnival Parade June 12. On the float is Ellen Smith, daughter of Attorney & Mrs. Hubert Smith, Glencoe. This float will appear in the Glencoe Fun Festival Parade July 10, as well as, on the occasion of surrounding celebrations.

The Co-op celebrated its twenty-fifth annual meeting and silver anniversary Thursday, February 25, 1960. About thirteen hundred people attended the meeting at Glencoe High School. Prior to the assembly more than thirty-one hundred members and visiting friends were served a noon luncheon at the Pla-Mor Ballroom. Some thirty ladies had started early in the morning preparing ham, beef, cheese, and wiener sandwiches, and pickles and doughnuts. The ladies also helped serve the lunch. Serving operations were under the direction of Bill Harpel and Lee Draeger, and fifty other members of the Glencoe Civic and Commerce Association.

At the business meeting, office manager Ed Reimler gave a financial report. Total revenue for the previous year was $815,407 of which $176,641 remained as net margin. The value of the entire electric system at the end of 1959 was $2,760,893. The Co-op had come a long way in twenty-five years.

Ancher Nelsen

On the Co-op's twenty-fifth anniversary, original Board member Ancher Nelson shared his thoughts:

It was my privilege to be on the board here in McLeod County, and later to become the National Administrator of this great REA program. In that capacity I was able to compare the progress and performance of the McLeod Cooperative Power Association with other systems. I was proud of the way our local Co-op measured up. Government payments were ahead of schedule, system improvements were well planned for future load growth, equipment was maintained perfectly, and it was obvious that there was an attitude of harmony in the business community, and among the board and all of the employees, which pointed to good management—the best.

It has been estimated, based on surveys, that the farmers of America have purchased about $15 billion of electrical equipment all over the United States. This has meant that the local merchant could sell to a new market. It has meant that the factories have had jobs for those who work in the plant. All of these things have added to the economic growth of the country, to say nothing of the tremendous benefits the farmers have received because of it.

GREEN GIANT AERATOR

We are happy to welcome the Green Giant Company as a new member-customer with the installation of the first of a series of aerators on their lagoons west of Glencoe.

It is hoped that this plan will enable the Green Giant Company to effectively cope with an air pollution problem that has plagued the vicinity at times. The Green Giant Company is to be commended for this forward look to becoming a better corporate citizen.

Shown above is a view of one of the lagoons partially filled and other under construction.

The above picture illustrates the size of the aerator. At the right is a huge paddle wheel to be operated by a 75 H.P. motor. Effluent from the canning plant operation will be preempt into this pit and aerated by the paddle wheel to accelerate air and oxygen action.

A power bank and service panel installed for the electric operation.

CHAPTER 6 1960s

Our Line's editor in 1960 noted that the Lakeside Store (located southwest of Hutchinson near Lake Allie) under the proprietorship of W. J. Kurth, had received service from the Cooperative since February 26, 1937. He was sorry that the business was closing as they were always helpful and cooperative. The store had been the central point of operations for that area during storms and line trouble. When telephones were out of service, members would leave their messages at the store and our men would make a point to stop there, and on those occasions Mr. Kurth always had a lunch ready for the men.

Water heater installations went up; 120 were added in 1960 and 125 in 1961. Time-controls, prevalent from the mid-1940s were considered counterproductive and removed by 1963.

In 1961, MCPA began to purchase and install automatically controlled mercury yard lights for members; initial monthly rental of $4.00 was later reduced to $1.50.

Yard lights provided a benefit on farm places, allowing more work to be done later and improving safety.

Electricity continued to show surprising versatility in the home and on the farm. MCPA played an innovative role in using off-peak stored electric heat. In 1962, Engineer Glenn Buck collaborated with Isakson Plumbing and Heating Company, Gibbon, to plan and install the area's first off-peak stored electric heat system in the Hutchinson home of Henry Baumetz. The system featured heating elements controlled by a time clock and peak control receiver; heat was stored in about seven hundred gallons of water.

STATISTICAL DATA

The following statistics based on January 1962 consumption proves interesting. Such gives a comparative analysis of each substation, indicating the capacity, showing the number of accounts, water heaters, kilowatt hours used, line loss, the percentage of total load used for water heating, average consumption, average amount used for electric house heating, and the load factor.

SUB-STATIONS	Accounts	Water Heaters	Total KWH Used	Total KWH Purchased	Line Loss %	KWH Used on WH	% Used on WH	Average KWH Per Acct.	Demand	E. Heat	L. Factor
Bell 3750 KVA	1151	865	1,086,180	1,178,712	92,532 7.8%	241,800	22.2%	943 KWH	2313 KW		68%
Bird Island 1500 KVA	739	481	652,170	705,024	52,854 7.4%	126,410	19.3%	882 KWH	1506 KW	3670 KWH	62%
N. Germany 2000 KVA	867	676	892,400	974,916	82,516 8.4%	188,160	21.0%	1029 KWH	2056 KW	3690 KWH	63%
Winthrop 2500 KVA	972	697	941,550	1,046,712	105,162 10%	171,020	18.1%	968 KWH	2166 KW N.A. 258	6130 KWH	64%
Hutchinson 1000 KVA	448	281	365,810	411,000	45,190 10.9%	73,080	19.9%	816 KWH	840 KW		65%

MCPA members added a variety of electric options: fireplaces, saunas, furnaces, and air conditioning. A 1967 newsletter noted an electric car-warmer to "pre-warm the inside of your car." An electric garden tractor was also marketed that year. By 1968, there were fifty "all-electric" homes in the MCPA project area and forty-five hundred consumers on MCPA lines. For them, electricity was a bargain—maybe not penny cheap—but close.

The capital credits program, launched late in 1957, was well underway. By 1961, credits were paid through 1950 and totaled $588,156. The program involved returning a share of margins, proportionate to the amount paid by a consumer over a year.

It was the year of trouble from Mother Nature in 1965. Those who were around at that time remember the worst snowstorm in many years, followed by spring flooding, and then a devastating tornado that swept through the Co-op's service area in early May. Over two-dozen farmsteads were severely damaged or completely annihilated.

The tornado swooped down about three miles south of Glencoe and tore through our new 69 kV transmission line in three places. Seven huge poles were broken off and a dozen uprooted completely. This knocked the High Island Substation in New Auburn Township completely out of service.

The tornado continued northeasterly, missing Glencoe by a quarter mile, and continuing towards Lester Prairie. In this twelve- to fourteen-mile path almost everything was demolished. Approximately ten miles of line was removed. Poles were broken and pulled out of the ground. Conductor was strewn in farm fields, meters destroyed, and transformers ripped from poles. One transformer was found in the crotch of a tree. Such destruction caused the Co-op a lot of trouble and expense.

Linemen worked all night and the next day. Help arrived in the form of a line contracting crew and a crew from the City of Hutchinson with their equipment, and several crews from Co-ops in Mankato, Litchfield, and Clements. Some co-ops offered materials and delivered it. MCPA had everyone back on that still had buildings standing within twenty-three hours. Many places had temporary lines that took thirty to forty days to replace back to normal construction standards.

Finally in 1967, Thiesfeld recalls, "wholesale power costs bottomed out at about six mills per kWh. Since then costs have been on a never-ending rise."

During the late 1960s the Cooperative still held its electric show the same day as its annual meeting. Area appliance and farm vendors could show their latest wares to consumers.

Line crew setting three-phase pole by Swan Lake in 1967.

The LeRoy Donnay farm was one of many hit by 1965 tornado.

Left: Not many persons are aware of our several warehouse buildings. The warehouse number 2 pictured above is located adjacent to the the south side of our main warehouse and office building in Glencoe.

Right: Warehouse number 3 was in the west part of Glencoe on our pole yard property beside the railroad tracks. We often have our poles delivered by train, and we then move them directly from the train cars to the pole piles.

Signs Placed - - -
(Continued from page 4)

Warehouse number 4 is located west of Glencoe on the road leading to Lake Marion from Glencoe. This is our most recently purchased building and is used to store some of our larger line hardware such as crossarms, wire reels, spare transformers and other like materials.

Waehouse number 5 is located in Hector for the crew serving our western area. This warehouse contains a small office, room for two trucks and storage for transformers, meters and line material necessary for day-to-day and emergency service work.

Four trucks were delivered to the cooperative in February to replace trucks which had received muchwear over the past two years. Trucks are replaced every two years before high mileage and wear causes excess maintenance expenses.

75 YEARS OF SERVICE TO THE MEMBERS OF THE MCLEOD COOPERATIVE POWER ASSOCIATION

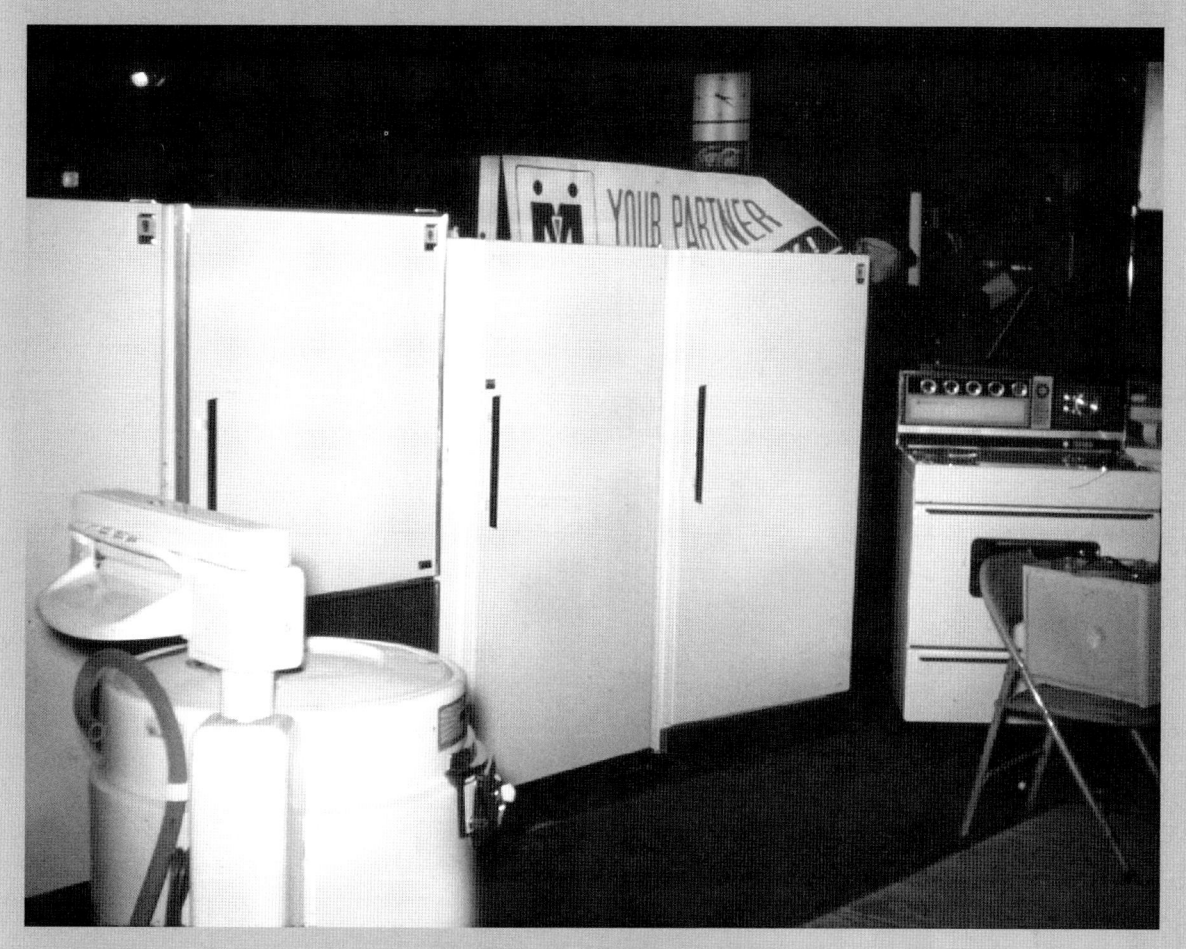

Local appliances dealers, such as the Midland Co-op booth shown here, gave consumers a look at the newest appliances for sale.

Carl Ortloff and manager Harry Theisfeld at the February 1968 electric show.

CHAPTER 6 1960s

CHAPTER 7
1970s

The annual meeting was always a big event for members, directors, and employees alike. For some members who lived further away, it may have been the only time during the year that they made a special trip to Glencoe. During all of the 1970s and until 1998 the Cooperative cleaned out its truck bays and served lunch to Co-op members right in the warehouse. The meeting would follow at the Pla-Mor Ballroom in Glencoe.

Left: Kentucky Fried Chicken was on the menu at the annual meeting. Employees and helpers packaged up meals for members.

Right: Members filled the truck bays and warehouse at MCPA's office to eat their lunch before the annual business meeting. For many years, the meal was served at the Co-op office.

In July of 1971, the Cooperative members no longer had to fill out their own electric bill. Members still read their meter but the Co-op began using an automated billing system. Members were now mailed their electric bills monthly. Statements were processed using a posting machine with magnetic stripe cards. It was not until 1985 that the Cooperative partnered with a data-processing group and began to enter readings and member's data electronically.

Following the low retail rates of the 1960s, a dramatic upturn took place during the mid-1970s. Orville Lipke, a Board member

OrDella Henke, seated at the terminal, along with Carol Barlau and Randy Owen, were becoming accustomed to new computerized data processing equipment in use at the Co-op. Over the years, many computer software and hardware upgrades have been implemented. The billing and accounting departments were responsible for implementing most of those changes, along with the data services cooperative providing us technical support, until recent years, when an internal software/technology person was employed full-time.

MCPA line workers set a pole in March of 1970.

Coal Creek Station generating plant near Underwood, North Dakota. The plant site is 2,560 acres and contains twin generating-units with a total capacity of one thousand MW. It is located next to Falkirk Mine. Coal is brought to the plant by a conveyor system. Not having to transport the coal long distances by rail or truck reduces the cost of producing electricity.

Fleet of trucks on pad in front of Co-op office. The Walkeracres building can be seen in the background.

at the time, cited contributing factors: the oil embargo, overall inflation, and higher-priced loans. Former Manager Thiesfeld believed the Eastern Seaboard power failure of 1965 led to increased governmental regulation. Compliance meant an added cost. At the same time, growth in member numbers and usage created demand exceeding "sources available to CPA," Thiesfeld said.

CPA had been purchasing power from Dairyland Cooperative, LaCrosse, Wisconsin; Upper Mississippi Valley Power Pool, and the U.S. Bureau of Reclamation. Looking for the "least expensive and most reliable option," CPA decided to build a mine-mouth power plant near the coal fields of North Dakota, Thiesfeld said. What ensued, he notes, was a period of frustration, not only for CPA but for each member cooperative as well. "The technology that had reduced power plant costs to $100 per kilowatt of capability, had to give way to inflation and bureaucracy increasing investment costs to well over $1,000 per kilowatt." These rising costs filtered into wholesale power costs for MCPA. "The cooperative was perplexed with inflation costs that had to be passed on to members in rising retail rates," Thiesfeld said. Rate hikes were obviously necessary. Increases were enacted in 1976 and 1978. In 1979, a 30 percent hike—believed to be the largest in MCPA history—went into effect.

Manager Ben Janowski explained the increase was "an entirely wholesale power increase" adding, "there

Opponents to the dc power line picketed and protested. Some vandalized the transmission towers being constructed, delaying progress and increasing the cost of the project.

Left to right: Members Ralph Novotny, Duane Kulberg, Vernon Ernst, Vernon Ruschmeyer, John Bernhagen, and Minnesota Governor Rudy Perpich, and Ray Wendtlandt who was present but is not seen in the picture, met to discuss the power line issue. The Co-op members presented the governor with a petition signed by members supporting the dc line on February 25, 1977.

is no doubt that any thought of 'cheap electricity' is gone. However, when used wisely and productively, electricity still does a lot of work for a reasonable fee."

After four years of construction, the first 500-megawatt unit of the Coal Creek generating plant went online August 1, 1979. Unit 2 came on in 1981. The plant, Falkirk Mine, the 400-KV direct current (dc) transmission line and the dc converter station, were a shared partnership between Cooperative Power Association (CPA) and United Power Association (UPA). CPA owned 56 percent and UPA owned 44 percent of the plant.

Orville Lipke, our own McLeod Cooperative Power Board member for twenty-nine years, served as the second Board president of CPA from 1968 to 1978. During these years, CPA and UPA were struggling to construct the dc power line from Underwood, North Dakota, to the converter station near Buffalo, Minnesota. Lipke was in the heart of the powerline controversy, organizing co-op members from across Minnesota to back the much-needed line that could provide Minnesota members with the power they needed to run their homes and farms.

A six-man delegation from MCPA presented a petition to Governor Perpich on February 25, 1977. Two of the men in the delegation, Ralph Novotny and Ray Wendtlandt, both had transmission lines running on their land and indicated they didn't consider it a drawback in purchasing a farm. They had worked under the lines the past twelve years.

Over thirty residents from McLeod Cooperative Power Association attended a pro-power line rally at the Capitol in St. Paul April 13, 1977, to voice support for the construction of the controversial direct current electric transmission in western Minnesota. Over twelve hundred persons from all over the state descended on the Capitol in what was described as the largest show of support for the line yet. Opposition

Placing the 1,657 towers along the 435-mile route required tremendous coordination. A change in the power line route, ordered by the Minnesota Environmental Quality Board, created much landowner unrest. Sixteen of the towers were toppled, and thousands of insulators shot out. The line was completed and energized October 17, 1978.

Farmer Orville Lipke was chosen to fill out the remaining term of District 5 Director Victor Hahn, when he died in 1956. Lipke was thirty-six years old and farmed in Round Grove Township. He ended up serving on the MCPA Board for twenty-nine years and was president of the CPA Board for ten years.

groups had only been able to muster two hundred people to protest the line's construction. Petitions with six thousand signatures in support of the line were left at the governor's office.

In 1977, Lipke was quoted in a press release as saying, "This line will be necessary if our cooperative is to have adequate electrical power for our needs in 1979. We feel that opposition to the building of the line, while it has been very vocal, expresses the feeling of only a small minority of people who just do not agree that the line is properly planned and needed. I can assure you it is."

Rosella Lipke, Orville's widow, reflected this past year on Orville's service to the Co-op: "In 1956, Orville was selected to fill the term of a member who died. He didn't want to serve, as he was busy farming and he knew how much time his dad had spent as a director, but he kept on and was reelected. One year he kept a tab and he spent seventy-two days of that year going to North Dakota when they were building Coal Creek Station."

New rates reflected operating costs of the station, the transmission line and coal purchases, as well as financing costs. Replacing insulators and other equipment damaged by vandals was an additional expense. Although rates were climbing, usage during the winter of 1979 topped the previous year by 15 percent. MCPA began to place more emphasis on energy savings through conservation and peak-trimming. As early as 1973, *Our Line* began featuring "conservatips" and focused on members' methods of energy conservation. For example, one issue focused on the John Bernhagen family's resolve to reduce energy consumption by 15 percent.

Left: Energy conservation became more important in the energy crisis of the 1970s and 1980s. Families became more conscious of their energy guzzling habits and some adjusted their appliance usage accordingly.

Right: Families would conserve energy more consciously as electricity prices increased. Thermostats were set back, water use conserved, lights turned off when rooms were not in use, etc. Microwave ovens came into use in kitchens, allowing for cooking with less energy use.

As wholesale power costs increased, the Cooperative tried to make consumers "less simultaneous" in their use, hoping to reduce generation and transmission costs. For this reason, Harry Thiesfeld explained, demand meters were installed in the early 1970s. "This was a very practical way of controlling peak-demand," Thiesfeld said. Demand meters were unpopular among many customers, however. Some felt they placed an unfair burden on large cash crop operations.

MCPA hosted its first Coal Creek Tour in the spring of 1979, before the first boiler unit was even operating. Thirty-nine members spent two days on a bus going to and from Bismarck, North Dakota, and touring the mine and power plant near Underwood, North Dakota. The group left at 9 a.m. on a Thursday and returned to Glencoe at 11 p.m. on Friday—a lot to see in a short amount of time. In the thirty years since this first tour, MCPA has taken its members almost annually to Coal Creek, except for those years that they could not fill the bus. Hundreds of Co-op members have seen first hand how ttheir power is generated.

CHAPTER 8
1980s

The off-peak water-heating program started in 1983. Each substation was outfitted with equipment to monitor loads. Information was transmitted by microwave to a computer at CPA's Eden Prairie headquarters, which tracked consumption through the entire system. When usage reached a pre-set level, CPA would send out the signal to drop the load. A receiver on the consumer's unit picked up the signal to drop or take up load.

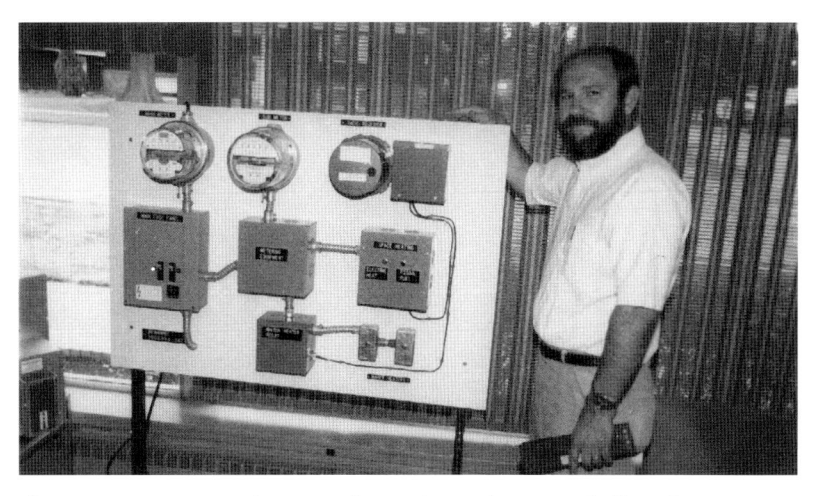

Left: Two fifty-two-gallon water heaters were the most common combination during the early years of off-peak storage water heating.

Right: Pete Malmin, engineer, showed a display board built to help members understand the various load management metering options.

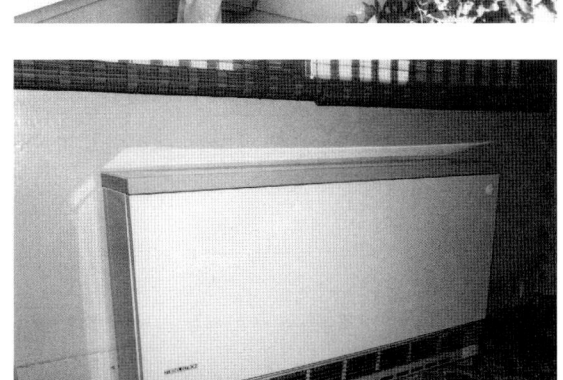

Top: Ron Sell tested a radio receiver and resealed a load management control box at a member's home.

Bottom: Room storage-heaters filled with ceramic bricks were installed by many members in the 1980s and 1990s. They operated on the storage-heating rate, using electricity to heat up the bricks only in the middle of the night.

About one thousand consumers on off-peak water-heating realize typical monthly savings of $4. Promoting load management programs for heating and water heating was the primary focus of the one-person Member Service Department during the 1980s. George Schwartz, member services manager, would educate members on the options available and then turn them over to their local electrician for installation of equipment. Employees from the metering and engineering areas assisted with checking out completed installations and maintaining radio receivers. Energy conservation education was also important during these years of high energy-costs and high inflation.

Room storage heaters were one of the popular units many members choose to install, often as a backup for their baseboards or other Dual Fuel electric heat. The system of two water heaters plumbed together for the Water Storage Program was also growing each year.

Left: MCPA line crews work to repair a broken transmission line pole south of Hutchinson along State Highway 15 on June 13, 1983. The storm knocked out power to 80 percent of MCPA members.

Right: The new warehouse, constructed in 1984, located along Highway 212 on the south side of Glencoe, is still in use today.

Office staff from the 1980s standing left to right, Nancy Weisert, Esther Beringer, Randy Owen, Mac Hoxie, Alan Drevlow, OrDella Knish, and seated, Sharon Maresh, Bernice Steinbrecher, and Mildred Graupmann.

A storm with damaging winds on June 13, 1983, caused widespread damage from Hector to Winsted. It broke off transmission line poles, putting 80 percent of the fifty-two hundred members out of power. Seven out of twelve substations were out of service.

Over the years, billing systems changed continually. As the number of billings increased, from perhaps 2,000 in 1941 to 5,241 in 1984, so did office staff. Mildred Graupmann, cashier, said that a punch-tape system was in place during the sixties. Information stored on the tape was shipped each month to a Twin Cities processing center. The method was replaced by an in-house ledger system.

In 1985, Graupmann said, data processing transmitted billing information to St. Louis, where statements were prepared and forwarded to Glencoe. Traditionally, the busiest seasons for MCPA office workers had been in the fall and in the spring, "when people moved." It entailed finding out who was moving out and who was moving in, getting the membership service agreement, and compiling a fact sheet for the computer, deactivating one account and activating another," she said.

From 1977 to 1986 the Co-op's newsletter was a few pages in the *Rural Minnesota News*, a monthly newspaper shared by many cooperatives. In 1986, we resumed our own publication called *McLeod Cooperative Power Association NEWS*, the name it still bears today. Sometimes between 1986 and 1994 our newsletter was inserted inside of the Cooperative Power magazine, published by our power supplier and mailed to our consumers.

Left: Cooperative employees celebrated 1 million safe working hours without an accident or major injury in January of 1986. They would continue to work without a lost-time accident until 1999, accruing well over 1.8 million safe working hours.

Right: Lineman changing out poles on a road job near Brownton in 1984.

Schatz Construction of Glencoe worked on the building in 1986, removing windows, improving insulation, and making the building more efficient.

After that time, just our MCPA *NEWS* was mailed directly to our members. In 2000, a larger color format was started, with some pages shared with Meeker Cooperative of Litchfield and Kandiyohi Power Cooperative of Willmar.

A momentous event was celebrated in 1986 when MCPA employees reached 1 million safe working hours. Monday, January 13, 1986, marked the 1 millionth hour worked without a disabling injury. Manager Ben Janowski cited several reasons for the Cooperative's outstanding safety record, including "a good training program, self-discipline of the employees, teamwork and a willingness to protect each other, and certainly over sixteen and a half years some good luck and fortune."

In 1986, the Cooperative's building underwent a facelift for energy conservation reasons. After thirty-six years, the old windows were removed and smaller, more efficient windows installed. Walls around the windows were insulated and faced with brick matching the original brick. The sleek, all-glass, modern look of 1950 gave way to the more practical, better-insulated style of the 1980s. Some painting and recarpeting were also done inside the building. And the engineering room was enlarged.

The other major undertaking of 1986 was director redistricting. Periodically, as the population of districts change, new director voting districts need to be formed to more fairly represent each area.

Annual meetings were always well-attended events right from the start. Members took ownership of their Cooperative and were loyal participants in the annual voter's meeting. Many years there were bands or some entertainment, music at the warehouse, good food, and lots of socializing.

Above: Entertainment was often a part of the Co-op's annual meeting celebration. Young musicians performed at the meeting in 1986.

Left: Mr. and Mrs. LeRoy Karg provided music for members eating in the warehouse.

Left: County dairy princesses helped with serving at the 1987 annual meeting.

Right: Pam Johnson, OrDella Henke, and Alan Drevlow worked annual meeting registration.

Left: Members of the pork producers cooked riblets in 1991.

Right: Members get served by lineman Curt Hanson at the 1999 annual meeting at the Pla-Mor Ballroom.

Members going through food line in the warehouse at 1987 annual meeting.

CHAPTER 8 1980s

Left: Mildred Graupmann and Alan Drevlow worked the registration desk at the annual meeting.

Right: Mark Walford, Don Jungclaus, and Ron Sell kept the coffee flowing.

Walkeracres building, which sat next to the Co-op's office to the south, was purchased by the Cooperative in 1988.

Excavation crews demolished the Walkeracres building the same year.

McLeod Cooperative Power purchased the old Walkeracres building, located south of the Co-op in May of 1988. The three-story structure was built in 1881 and served as a hotel for many years. In 1928, Charles Walker purchased the building and used it as a hatchery. His son, Charles Walker, purchased the business in 1953 and operated it as Walkeracres until 1985. The last three years of its life, it was used for storage. The wooden structure could have been a substantial hazard to the Co-op if it ever caught fire. The Co-op had the building demolished during the summer of 1988.

An aerial photo of Co-op office in 1986.

Part of the truck fleet in 1988.

Left: Ninth grade science class from Brownton visited the Co-op for a safety demonstration in 1989.

Below: Co-op personnel worked with students at Glencoe High School in 1990 on electrical safety. The Cooperative continues to present safety education to students today.

CHAPTER 8 1980s

Left: Operations and engineering employees participate in 1985 safety training session. Safety training has been a critical component in maintaining the Co-op's excellent safety record year after year.

Right: Superintendent Mark Walford and Don Jungclaus (seated) use the system map to pinpoint an outage in 1985.

Members stopped in for coffee, cider, and cookies at the 1982 Christmas Open House.

An unplanned power interruption began on November 18, 1986, when Unit No. 2 boiler at Coal Creek Station power plant exploded. No one was injured but damage was extensive and repairs took many months. Power purchases had to be made from other utilities to replace a large portion of the electric generation for an extended period.

The Cooperative began supporting ethanol long before it served any ethanol plants. In 1988, it began using a 10 percent blend in the trucks. By the mid-1990s the Co-op started burning E-85, an 85 percent ethanol blend fuel, in some smaller vehicles and passenger cars. Most of the pickup trucks and passenger vehicles at the Cooperative today operate on E-85.

For decades the Cooperative has been educating school students and the public on working and living safely around electricity. In the 1980s the Cooperative was hosting school groups or visiting schools with their electric safety message. In 2009, Cooperative employees are still visiting schools regularly with a similar safety message.

CHAPTER 9
1990s

The early 1990s were a difficult time for MCPA management, as they had to deal with the erosion of cooperative service territory by annexation. The City of Hutchinson took service territory from MCPA along Otter Lake, bare land that would soon be a large housing development west of Hutchinson, the current county fairgrounds land, several churches, and most of the land that today is home to Wal-Mart, Menards, and the other commercial real estate on the southwest side of Hutchinson. Although the City had to provide some monetary compensation to the Cooperative, the Association did not feel it was near enough to give up rights to those developing areas that had been its service territory since 1935. It was a time of difficult negotiations and strained relations with the City of Hutchinson.

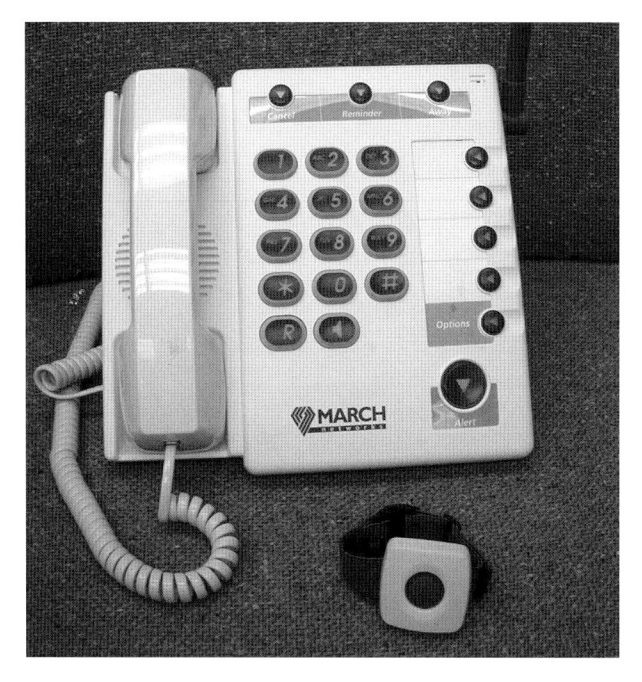

Above: Cooperative linemen trench in cable. Primary underground cable and secondary cable were becoming more common installations. Although usually more expensive, URG cable had advantages on certain job sites where it was difficult to put in poles or if for safety reasons underground was needed.

Top left: Line superintendent Mark Walford checks fuses in a disconnect box.

Bottom left: MainStreet Messenger telephones and emergency pendants have been used by many members and nonmembers alike since the Co-op began offering service in 1993.

McLeod Cooperative Power Association, along with eighteen other rural electric cooperatives in the Upper Midwest, formed Cooperative Response Center (CRC) located in Austin, Minnesota, in 1992. CRC was founded to meet the after-hours dispatching needs of those cooperatives that formed it. It was becoming increasingly important to have a real live person answering the phone when our members called on evenings or weekends, whether it was an outage or they needed other help from the Co-op. CRC met this dispatching need plus they also provided monitoring service for MainStreet Messenger emergency medical systems.

DIRECTV systems were new in 1994. Consumers had never experienced satellite service over a small dish before. Sales were fast and furious, with hundreds of members on a waiting list for equipment as it became available. The first consumers to get equipment in those early years had cheaper programming but it cost them $699 to $899 for a dish and receiver system.

That led the Cooperative right into a new venture, that of providing professionally monitored medical phones for senior citizens and disabled persons. The business of providing this community service has continued for sixteen years. It was a worthwhile service that benefited many families, even if it only produced marginal income for the Cooperative. It was one more thing we could offer to both our members and nonmembers alike.

The Cooperative directors returned from a National Rural Electric Cooperative Association (NRECA) meeting with information that satellite TV was to be the next up-and-coming innovation. At that time there was cable TV if you lived in a larger town or city. There was no pay-TV in the rural area unless you bought yourself one of those huge satellite dishes.

NRTC, the National Rural Telecommunications Cooperative, had formed, and was going to be offering the opportunity for Co-ops to invest in this business and provide the service in their individual service territories. DirecTV at the time was a fledgling corporation in need of financial partners to help fund the launch of its first satellite into space. The Cooperatives had the money to invest and had the existing customer base and infrastructure to promote the product. Never did we guess that satellite TV would grow into such a huge industry or DIRECTV into such a giant of a company (so big that they now use all capital letters in their name). We never would have believed in 1992 how this business would eventually impact McLeod Cooperative Power.

So, in November of 1992, at the request of the Board of Directors, a survey was sent to all of the electric members, asking if they would be interested in satellite TV, if the Co-op were to offer it. Over seventeen hundred electric members returned their surveys, indicating more than 52 percent were interested. That was the answer the Board needed to proceed. At the time, the Board of Directors and management thought it was a huge amount of money to buy into the program. For us to be the sole DirecTV provider for all of McLeod, Renville, and Sibley Counties, plus all of the rural homes in western Carver County cost us $415,610.15. The Co-op contemplated purchasing all the homes in Carver County, but could not afford it because of the large population in the eastern end of Carver County.

"At the time it was a lot of money," reflected Lester Ranzau, then president of the Board. Satellite TV off of a small dish was a service that in 1992 was not available from any provider. It was new technology that was not yet on the market. According to President Ranzau, the Board's decision came down to, "It was a service for our members, that they were not receiving, and we could provide it." The Board voted to move forward and complete the purchase of rights to provide DirecTV to homes in the counties and zip code areas the Co-op had specified.

Then the waiting started. Waiting for the technology to be developed. Waiting for the first satellite to launch successfully. The Co-op waited for RCA to start manufacturing receiver boxes and equipment to for them to sell. During this waiting time the Cooperative hired Bob Thomes, a local fellow with experience in TV and electronics sales, to help get this new retail sales venture going.

The first satellite went into orbit successfully in March of 1994 and the second was deployed that summer. In April of 1994, the Co-op started taking reservations for Direct Broadcast Satellite (DBS) equipment. Cooperative members lined up to get their names on that waiting list, even though equipment prices were $699, $799, and $899 initially. By July there were over a hundred members on a waiting list for equipment. June 17, 1994, DirecTV service aired, although our customers did not yet have receivers. The customers who were first on the waiting list started receiving their equipment in early August.

Above: The Cooperative was a county fair exhibitor in the 1940s and 1950s when new electric appliances were all the rage. The Co-op also exhibited at McLeod, Renville, Sibley, and Carver County Fairs from about 1993 to 2004, during the busy DIRECTV years. This was our booth at the Renville County Fair in Bird Island in the late 1990s.

Right: It took four Co-op employees to operate the DIRECTV business for many years. Front row, Katie Ide and Deb Goettl; back row, Bob Thomes and Debbie Ebert.

The Cooperative had a working unit that it took on the road to its four county fairs. It was quite the attention-getter. Everyone at the fairs had to stop and watch or ask about the details of DirecTV, at that time the only provider of any small-dish satellite TV service. It was something new and wonderful that people had not seen before! The Co-op continued working the county fairs until 2004.

Programming packages were small by today's standards but so were prices. Customers received seventeen channels for $16.95 a month in the Economy Basic package or thirty-eight channels plus music channels for $29.95 per month in Total Choice.

In July of 1995 the Co-op added a toll-free telephone line to accommodate all of those DIRECTV customers calling from outside of the Glencoe calling area. DIRECTV just kept growing. By 1996 the franchise had 1,400 customers. By 1997 there were 2,049 subscribers—a 44 percent increase in one year. This growth accounted for

OrDella Knish hands out capital credit checks at the Co-op's holiday open house.

Randall Owen, MCPA general manager, gives an energy grant check to Bill Adcock, Heartland Corn Products manager. Heartland received funds for implementing energy efficient motors and equipment in the plant to reduce peak demand.

the addition of some new employees. At the peak of DIRECTV sales, four employees were working in that area, plus the Co-op utilized the contract services of several local installers.

McLeod Cooperative Power started a partnership with Ridgewater College Foundation in 1993. Since then the Cooperative has supported scholarships for students at Ridgewater College through donations of unclaimed capital credit funds to the Ridgewater Foundation.

The mid-1990s brought huge growth to the electric business as well. In 1994, Heartland Corn Products in Winthrop, Minnesota, began producing ethanol. It started as a 4-million-gallon-per-year plant. This load was shared between McLeod Cooperative Power and Brown County REA of Sleepy Eye. The Co-ops shared in the investment and the risk, as well as the revenue from this constantly growing and advancing ethanol plant. In 1995 was the groundbreaking for Minnesota Energy, an ethanol plant in Buffalo Lake, Minnesota. Today, Heartland Corn Products has grown to be a 100-million-gallon-per-year plant, and Minnesota Energy produces 19 million gallons per year plus operates an elevator, grain, and fuels business.

The Cooperative continued on its run of 1.6 million safe-working hours by its employees. This was one of the best, if not *the* best, safety record of any utility in the nation. The Co-op was honored by the Minnesota Safety

Minnesota Energy ethanol plant in Buffalo Lake.

CHAPTER 9 1990s

Heartland Corn Products ethanol plant in Winthrop.

MCPA linemen use the Co-op's Bombardier with tracks to dig a hole and set a pole north of Silver Lake.

Members participated in one of the Cooperative's first wind farm tours. The large tour bus is dwarfed by the large wind tower.

Council in 1996 for the 1,648,723 hours its employees had worked without a lost-time accident. The Cooperative's accident-free period lasted from July of 1969 until July of 1999, with over 1.8 million work hours accrued.

Annexation of Cooperative service territory happened again when the City of Hutchinson chose to annex twenty-four existing McLeod Cooperative Power customers. The City also annexed bare land, scheduled to eventually develop into hundreds of residential housing units. These annexation issues have strained relations between the Cooperative and the City who historically before the 1990s had a good track record of working together.

Renewable wind energy was something new beginning in 1997 that the CPA cooperatives made available to their customers. The Cooperative offered to members the purchase of wind energy for a premium amount of

Winsted on the Lake housing development near Winsted was the Co-op's first large development.

$4.00 per 100 kWh per month. Once enough commitments were made from CPA member co-ops to purchase all the wind generated by Chandler Hills Wind Farm, construction began on the turbines. Power generated was replacing energy generated by the same amount of fossil fuel. It was member's first chance to show their support for green power, if they were so inclined. Eventually, as the cost of building and producing wind decreased, the premium amount member's pay had dropped to $1.50 per 100 kWh per month.

The Cooperative had its first very large housing development begin in 1997. Scenic Homes, better known as Winsted on the Lake, was a 157-unit housing development on the east side of Winsted, which included single family, twin, and town homes. It continued to expand for several years until all of the lots were sold. Cooperative personnel worked with the City of Winsted on plans for the development, as well as with the developer to include off-peak water heating and air conditioning in the homes being built. Other development followed, such as Prairie Ridge in Lester Prairie and Hidden Creek in Mayer. Homes were quickly added to each of these developments until the recession of 2008–2009 hit, and sales stalled and they did not complete houses in the final phases of the development.

Excavation began in 1998 on an addition to the west side of the Co-op office.

MCPA along with hundreds of other cooperatives across the nation formed Touchstone Energy, a brand designed to position cooperatives for possible deregulation. MCPA continued with this marketing group for many years, until the threat of deregulation had faded from the difficulties in states that had tried deregulation. The Cooperative marketed this Touchstone message with other Minnesota cooperatives at Farmfest, where they all had a booth, as well as at its own functions. And they were part-sponsors of the famous Touchstone TV ads and commercial jingle.

The Cooperative undertook a building expansion project in June of 1998 and added twenty-two hundred square feet to the west side of the building. It gave the Co-op a large meeting room that would accommodate all of the employees at one time, a new lunchroom, and a new engineering room. The old lunchroom and engineering room were renovated into office space for other departments. A ribbon-cutting ceremony was held December 16, 1998.

Left: Block work was done for the exterior wall of the new board room, kitchen, and engineering room.

Right: Finished exterior of backside of MCPA office is shown. The addition includes under-floor electric heat.

At the ribbon cutting in December 1998 for the new addition, members of the Glencoe Chamber of Commerce, Co-op Board of Directors, and Manager Randall F. Owen, officially celebrated the completion of the addition.

The new meeting room was decorated for Christmas.

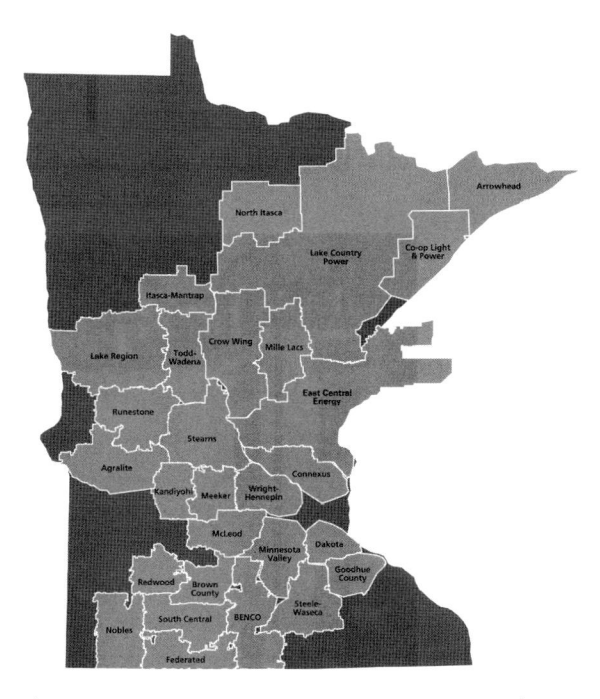

Above: Great River Energy service area map. It shows the distribution cooperatives that are members of the newly formed GRE.

Above right: Heartland Security Services, with its fleet of installer/technicians, are available to provide reasonably priced, but professional, security monitoring for homes, businesses, and farms within our service area.

Line crews installed underground service to a farm west of Winsted.

MCPA's power supplier, Cooperative Power Association, headquartered in Eden Prairie, merged with United Power Association, Elk River, on January 1, 1999. This formed Minnesota's largest generation and transmission (G&T) cooperative, named Great River Energy. It was a natural move as the two cooperatives were both owners of Coal Creek Station and the dc power line and other facilities. It was hoped that this merger would result in great financial savings for the member cooperatives, although end results were hard to measure.

Another new business formed in 1999 was Heartland Security Services. Eight rural electric cooperatives from Minnesota and Iowa formed Heartland Security Services to offer another diversified service to their members. Heartland has continued to grow to ownership by thirteen rural electric cooperatives, has multiplied its number of subscribers several times, and has become a financially viable business. It is just one more service the Co-op can provide to its members at a reasonable rate and can guarantee them a high quality of service.

The summer of 1999 had some record-setting high temperatures and record-setting high energy prices. MCPA's power supplier, Great River Energy, exceeded 1,900 MW peak demand on July 30. GRE had to pay premium prices for electricity it purchased on the market to keep up with consumer demand. The effect of deregulation in some parts of the country played a part in skyrocketing prices in the wholesale electricity market. Cooperative members felt the bite of those high prices passed along on energy bills. MCPA members were very helpful in their conservation efforts that summer and participants in load management programs experienced more control hours than usual.

The number of construction projects completed in 1999 was a record for recent years. Although nothing like the 1937–1947 era, it definitely kept the crews busy. Eighty new services were built and eighty-one service rebuilds were completed. The housing boom has been the main driver for this activity.

With year 2000 looming, utilities all over the world had to prepare for Y2K, uncertain whether any system would fail or not. The Cooperative, like most utilities, prepared by testing its systems and doing drills and working with its vendors to make sure equipment it used would not be affected adversely. Fortunately, it was a lot of preparation for operational problems that never actually occurred. Co-op personnel were even on hand through the middle of the night on December 31, 1999, just to make sure they were available to handle any events that might have happened. All of the Co-op's computer and electrical systems continued without interruption.

Line foreman Mike Gassman placing jumpers on a line.

Directors Curtis Rossow, Dale Peters, Gerald Roepke, Lester Ranzau, Mel Burns, Roger Karstens, and Charles Jensen were at the Minnesota Capitol rotunda for Technologies Day, featuring Richcraft Wood Products display, and also to lobby legislators.

CHAPTER 10
2000 TO 2009

MCPA began providing maintenance and outage services to the City of Arlington in 2000 on a contract basis. Before that time it was just on a trial basis.

The Cooperative's old logo was used for decades. It is shown here in a display with broken insulators.

The Co-op's new logo was rolled out in 2001. It more clearly shows the variety of services the Co-op provides.

McLeod Cooperative Power Association started out the new millennium with a new contract customer. The City of Arlington signed a contract with MCPA to provide its maintenance and service work. The Co-op had been assisting the City for several months on a trial basis. All went well and on January 1, 2000, MCPA began working under the contract to respond to their outages, do the City's electrical service work, and some engineering services. This added an additional one thousand electric services that the Co-op would be responsible for keeping energized, however, it was viewed as a win-win situation for both parties. It improved service response for consumers and met financial goals of both organizations. The City of Arlington continued to do its own billing, underground locating, and tree trimming. In 2009, the City and the Cooperative are still working together under this same arrangement.

There were record kilowatt-hour sales at MCPA in the year 2000. Members purchased over 140 million kWh. DIRECTV also posted a profit of $129,000 and had 4,320 subscribers at the end of the 2000. As of April 1, 2001, that number had already passed 4,500.

The Cooperative designed and unveiled a new logo in January of 2001 to replace the old-style logo in use for many decades. The new logo was designed to clearly show the Co-op's name and full range of services offered to rural residents. The green and blue colors symbolized MCPA's commitment to the environment and renewable energy.

The residents around Lake Allie, north of Buffalo Lake, formed an environmental subordinate service district in 2001 and built a wastewater handling system for all Co-op members living around the lake. The system solved the problem of lake residents with failing sewer systems and gave them an environmentally acceptable system to use into the future that will help preserve the quality of their lake water. The Cooperative is working with the Renville County service district by doing the monthly billing to members for their wastewater services. This saves them money, as they are not paying so much for postage or sending a separate bill. This system has worked fine for both parties, and is still in use today.

Left: The front office of the Cooperative was under construction in 2003.

Right: With remodeling complete the front office had more room for products such as our DIRECTV display.

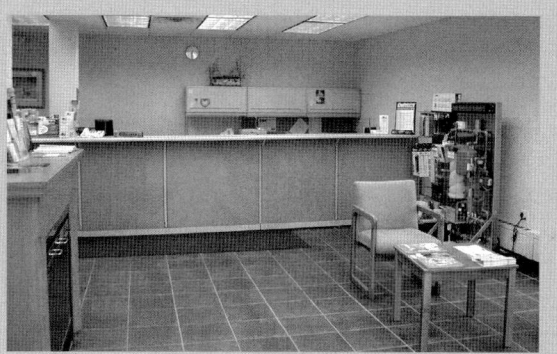

Left: There was now room for a Marathon water heater and fireplace in the lobby.

Right: Two counters were now available to members for paying DIRECTV or electric bills and getting customer assistance.

From the outside, the finished addition replaced the planter with building square footage. The Co-op has a handicapped access on the level on the west side of the building.

CHAPTER 10 2000 TO 2009

The gals in the front office decorated the office for the holidays capturing first place in the Glencoe Chamber of Commerce decorating contest and winning a large sub sandwich for the office.

This Sibley County youth group received funding from Operation Round Up.

The main lobby of the Cooperative's headquarters building received a facelift in 2003. After the updating by contractor Schatz Construction, the Co-op had more room to display the products and services it offered. Changes also included making the entry safer. The twenty-four-hour payment drop box was moved to sidewalk level, eliminating the need for consumers to climb stairs to access the drop box. Electric heating panels installed in the sidewalk area, just outside the front door, provide snow and ice melting in inclement weather.

One of the biggest projects undertaken by the Cooperative in many years was the deployment of a Turtle automated meter-reading system. Purchased from Hunt Technologies, the system will report each member's monthly meter readings to the office, eliminating the member or Co-op employees from having to read it each month. The deployment of Turtles and new meters began in 2004 and single-phase meters were finished in February of 2006. Three-phase meters were completed that spring.

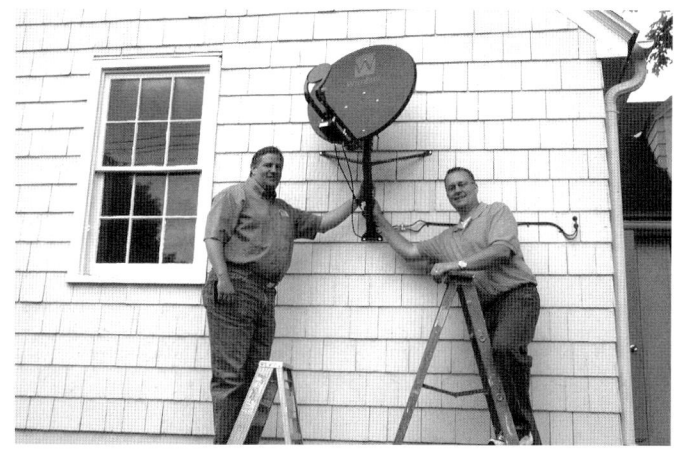

The Cooperative began installing WildBlue high-speed Internet in 2005.

An Operation Round Up program was started in 2004. The program allows members to round up their electric bill to the nearest dollar, and that change will go into a donation fund. It is overseen by a board of five volunteer persons appointed by the Board of Directors and they award the funds available each year to applying charities and special organization projects in our service area. From 2005 to 2009, MCPA's Operation Round Up fund has provided $14,200 to help fund forty-two projects in the Co-op's four-county area.

In 2004, DIRECTV offered to purchase the Cooperative's exclusive rights to provide programming to McLeod, Renville, Sibley, and parts of Carver Counties for $5.5 million. This purchase price would be paid to the Cooperative over seven years with interest payments, bringing the total to about $7 million. Although the entire $5.5 million transaction had to be shown on the Cooperative's financial books for 2004, the Cooperative would actually receive only an equal portion of that each year, with interest, for the seven-year payment period. This was a good return on the Co-op's original $415,000 investment.

At this same time, the Cooperative entered into an agreement with DIRECTV, authorizing MCPA to be a retailer, sales agent, and billing/programming provider. This made it seamless for the customer. They could continue to call the Co-op twenty-four hours a day for DIRECTV service, make DIRECTV payments or purchase equipment at our Glencoe office. Although the Cooperative would no longer have control over pricing or administrative decisions, the daily operation of the DIRECTV business would change very little for consumers.

Financially, the DIRECTV business provided significant nonoperating margins to the Cooperative's bottom line for several years. This helped the Co-op keep rates stable at a time when wholesale energy rates were climbing. MCPA's Board of Directors developed a plan to utilize the dollars of the DIRECTV sale for the long-term benefit of the Cooperative and all of its electric members. The Co-op also had to make certain that the plan abided by Rural Utility Services (RUS) accounting guidelines and all tax laws that limit some of the uses of these funds. Some of the revenue from the sale has been used to finance electric construction in lieu of borrowing funds from RUS or commercial lenders. A portion of the funds were used for business ventures like WildBlue high-speed Internet, DIRECTV service, and funding deployment of the Turtle meter-reading system.

The Main Street Market and Down Home Bakery opened on the Main Street of Buffalo Lake. The building was partly financed by a $300,000 USDA grant the Co-op helped secure.

New visual computer equipment in use by employees Greg Nistler, Dave Keil, and Eric Sell in the engineering room.

Annual meeting crowd in 2003.

In 2005 the Co-op began offering WildBlue high-speed Internet service. It is a satellite-based signal and requires only a small dish in the yard or on the house. No phone line is required and it is available for rural residents who want speed faster than dial up but do not have access to DSL. As of 2009, the Cooperative had approximately five hundred subscribers using the service.

In 2005, things were shaping up in downtown Buffalo Lake. Following a tornado two years previous, the downtown area had been devastated and some of the main businesses in town were destroyed. Although there was nearly $14 million in damage, the City did not qualify for FEMA assistance. Local residents did not qualify for low-interest loans. They had to pull themselves up by their bootstraps.

With a little help from Renville County EDA, USDA, McLeod Cooperative Power, their own Welding Fund, and CenBank, they were able to rebuild their Main Street. The Co-op's part in the project was securing a $300,000 grant from USDA to fund part of the building project. The Buffalo Lake EDA (Economic Development Authority) took on the project of building a new bakery and grocery store. Other businesses were repaired. By 2005, when the bakery and grocery store opened, Buffalo Lake was looking even better than before. Between 2005 and today, the residents of Buffalo Lake have continued to improve and grow their community.

For the Operations and Engineering Departments, new technology has been one of the major behind-the-scenes changes to improve the efficiency of the work stakers and linemen perform. Sometimes it has been an adjustment for employees to learn new ways to do the same old job using computer technology instead of writing it all down on paper. But eventually, the systems will be integrated to do more work automatically, telling accounting what materials have been used on each job and posting them to work orders. Mapping is one area that is now easier and faster for Co-op employees to update electronically. No more waiting years for a new map book to be updated and printed.

Left: Annual meeting crowd at Hutchinson Event Center in 2009.

Right: Head table at the 2009 annual meeting, where Co-op employees showed a slide presentation.

The Cooperative purchased the corner lot southeast of the Co-op in 2009 from the City of Glencoe, formerly known as the Bruckshen property.

New home construction and load management system sales kept MCPA staff especially busy from 2000 until the recession slowed things down in 2008.

In 2009 the Cooperative moved its annual meeting to the Hutchinson Event Center. It proved to be a good move, especially for the older patrons, who could more easily get around in the handicapped accessible building. About 560 members attended the business meeting, followed by a turkey dinner for lunch.

Late in 2009, the Cooperative purchased the Bruckschen property, directly to the southeast of the Cooperative's office. It is a corner lot that will allow for future expansion and improved off-street parking for employees and patrons. No exact plans have been developed for the property. The Cooperative was interested in the parcel for a long time, even before the building on the site was torn down and the site cleaned up, but it took until 2009 for the City of Glencoe and the Co-op to agree to terms. The City had purchased it several years earlier as part of an economic development project. The Cooperative also purchased the former Agri-Fleet building directly east of the main office.

Meter technician Bob Senst shows member Al Huff how to read his new electronic meter. Al has a wind turbine co-generating power back to the Cooperative.

In the last few years, the Cooperative has been spending considerably more time on energy conservation and renewable energy programs. This has happened because of consumer interest, government mandates and tough financial times for consumers. Our power supplier, GRE, has met most of our renewable energy requirements through wind farms like Trimont, Prairie Star, and other renewable biomass projects. Locally, the Cooperative is working with an ever-increasing number of members installing small wind turbines (those under 40 kW) that produce power for their own need and sometimes have excess generation to sell back to the Cooperative. In 2009, we had about half a dozen projects operating.

Cost of service for our members has always been important to us. This year and probably in the coming years, federal and state regulation will affect the cost of electricity and how fast that cost increases, more than anything else. That is why the MCPA Board of Directors and management are diligent in monitoring legislation, and working with both state and federal elected officials, to protect our members from unnecessary increases.

Linemen Dan Schade and Craig Marti install underground conductor to a new URG transformer just north of Glencoe.

Photo on opposite page: Orville Lipke, on left, turned the gavel over to new Board president Lester Ranzau in 1983. Lipke served thirteen years as MCPA Board president and Ranzau served twenty-two years as president.

CHAPTER 11
GENERAL MANAGERS AND BOARD OF DIRECTORS

R. A. (Ben) Fischer, Manager 1939 to 1966

As the county agent, Ben Fischer worked tirelessly to mobilize the McLeod Cooperative Power Association and get it established from 1935 to 1939. In 1939 the Board extended an invitation to Fischer to serve as general manager of the new cooperative. In the Co-op's twenty-fifth anniversary history, Fischer recalls his reaction: "When asked to assume the position as manager of this Association, such came as a surprise. Even though I had been working with those in charge I had given no thought of becoming thus associated with it. My first reaction was to say no. Upon being advised that it was the unanimous request of the Board of Directors with whom I had worked so closely, the decision became more difficult, and after discussing it with my wife, we decided to launch upon a new adventure. The work has been challenging, adventuresome, and interesting. We in the Association went through the years of World War II, and during that period built lines continuously, so as to serve our people."

Ben Fishcher

Former Board member Wayne Bulau described Fischer as having grown with the association. "When he started, he didn't know an insulator from a kilowatt. He learned from the school of hard knocks. He was a good man," said Bulau.

Fischer's hard work and dedication earned the respect of the Board and the trust of the members." He was a tremendous manager," former Board member Lynn Wulkan remarked. As a longtime resident of the area, he knew most of the farmers well and had many contacts when he needed volunteers.

A native of Cavalier, North Dakota, Fischer was raised on a farm north of Buffalo Lake, Minnesota, and graduated from Hector High School. He was a graduate of the University of Minnesota College of Agriculture and served as a county agent after graduation.

Fischer retired from the Co-op in 1966 and died in 1979.

Harry W. Thiesfeld, Manager 1966 to 1977

Harry W. Thiesfeld of Glencoe joined McLeod Cooperative Power in 1963 as engineer and assistant manager. In 1966, he succeeded Ben Fischer as general manager. He had been employed by the Rural Electrical Administration for twenty-five years as a power planning engineer, before coming to McLeod Cooperative Power. His thorough knowledge of REA operations was a valuable asset to the Cooperative. Thiesfeld said he welcomed the change of working with people directly at MCPA after years of bureaucracy.

Thiesfeld worked to trim peak power requirements for the Cooperative and to improve load management programs. He was active in statewide associations.

In 1977, Thiesfeld retired to join the Electric Guidance Team in Bangladesh.

Harry Theisfeld

Bernard (Ben) Janowski, Manager 1977 to 1991

Ben Janowski became the third manager of the Cooperative on April 1, 1977. He came to Glencoe from Cando, North Dakota, where he had been the business manager for Baker Electric Cooperative. He served in many civic and business organizations in the Glencoe community while he served as manager.

Janowski's leadership kept McLeod Cooperative Power on a sound basis during times of high inflation, high construction costs, and rising rates.

Ben Janowski

Before coming to Glencoe, Janowski attended Lake Region Community College in Devil's Lake, North Dakota, and he served a National Rural Electric Cooperative Association (NRECA) internship at the University of Nebraska. He spent ten years working for the State of North Dakota and twenty-eight years with the rural electric cooperatives, before retiring in 1991. He and his wife, Marion, now split their time each year between Minnesota and Arizona.

Randall F. (Randy) Owen, Manager 1991 to 2007

Randy Owen was employed at McLeod Co-op Power for thirty-six years. He started with the Co-op as an accountant in 1971 and was promoted to office manager in 1978. Randy served as assistant general manager from 1989 to 1990 and took over as general manager in 1991.

During Randy's fifteen years as general manager, the Cooperative built two additions to the headquarters building and an additional building at the pole yard on Highway 212. A major undertaking for the Cooperative was buying into the DIRECTV business in 1993. This business grew to serve five thousand subscriber homes in a four-county area. He saw the cooperative through Y2K preparations, deployment of Turtle automated meter reading for our entire project area, and the start-up of an Operation Round Up program that makes contributions to local charities and organizations. In 2000, Randy also reached an agreement with the City of Arlington to provide electrical maintenance and engineering services to the City.

Randall F. Owen

CHAPTER 11 GENERAL MANAGERS AND BOARD DIRECTORS

Randy and his wife Faye lived in Glencoe and raised their family there for thirty-six years. Their retirement includes more time with their children and grandchildren, as well as golfing, fishing, and traveling. They reside near Brainerd, Minnesota.

Kris Ingenthron, Manager 2007 to date

Kris Ingenthron, current manager for McLeod Cooperative Power Association began his career in 1980 as a one-thousand-hour employee with an electric cooperative in Wisconsin. In 1985, Kris moved to Montana, working as a groundman for a power line contractor. Ingenthron became a journeymen lineman in 1996 while employed with Sun River Electric Cooperative, Fairfield, Montana. In 1998, Ingenthron accepted a position with Oconto Electric Cooperative, Oconto Falls, Wisconsin, as their line superintendent. During his tenure at OEC Kris advanced to vice-president of operations as well as being appointed interim general manager during the executive search for the cooperative. Ingenthron is a graduate of the National Rural Electric Cooperative Association (NRECA) Robert I. Kabat "Management Internship Program."

While Ingenthron's tenure at MCPA is relatively short, he has overseen the conversion of the customer information system, the accounting system, and the electronic staking/mapping system, and continues to advocate advancement in the technology side of the cooperative.

Kris Ingenthron

While the electric utility business continues to change, Ingenthron's goal is for the cooperative to continue focusing on the core business while still keeping one eye open for future business opportunities. With the employee average age increasing, the cooperative will need to address organizational restructuring along with looking at the future building needs of the cooperative.

Kris and his wife, Collette, have two children, Jared (Tiffany) of Oconto Falls, Wisconsin, and Tucker, a sophomore at Glencoe-Silver Lake High School. Kris and Collette are also proud grandparents to grandson, Jack. They spend their free time following their son Tucker's soccer program along with fishing, boating, volunteering, and quality family time.

Board of Directors

It is with great appreciation that the Cooperative honors all the current and past Board members, who have given of their personal time to advance the growth of McLeod Cooperative Power Association. It is our directors who represent all of the electric members in their district and the membership as a whole, making decisions to best serve the needs of the membership.

Lars Leifson, President 1935 to 1943

Lars Leifson, known as the "grand man" of MCPA, made a motion to organize an electric cooperative in 1935, and was promptly named its first chairman. At the first annual meeting in 1936, Leifson became MCPA president, a post he held until retiring from farming in 1943.

Born in Mandol, Norway, Leifson immigrated to the United States at the age of eighteen. He worked to pay back his borrowed passage and in eight years, had even saved enough to buy an eighty-acre farm, six miles northeast of Glencoe.

An early advocate of the benefits of electricity, Leifson was quick to install a farm-generating plant that was operational on the farm for nineteen years. Leifson's farm was one of the first to be energized on June 10, 1937. Lars Leifson is the proud possessor of Stock Certificate No. 1 in the McLeod Cooperative Power Association, and that is as it should be. He was the No. 1 booster, worker, and promoter of our project, and a great deal of credit must be given him for its outstanding success. The family soon added a full complement of electrical appliances.

After moving to Glencoe, Leifson continued his MCPA affiliation assisting in rights-of-way negotiations and staking operations. When the new building was completed, Leifson served as custodian until retiring in 1957. At that time, former manager Ben Fischer wrote in Our Line: "Mr. Leifson always took a deep interest in the affairs of this Association, which, aside from his family, is virtually his first love. He led and counseled wisely. He truly is the "Grand Man" of the association. Leifson died October 7, 1966 at the age of eighty-four.

Directors and Their Tenure

Names and Offices	Tenure
Lars Leifson President 1935 to 1943	1935 to 1943
Herman Graupmann Vice President 1935 to 1943 President 1943 to 1948	1935 to 1948
Walter Jungclaus Secretary 1935 to 1940 Vice President 1950 to 1951	1935 to 1951
John C. Lipke Treasurer 1935 to 1939 Vice President 1948 to 1950	1935 to 1950
Harry Tews	1935 to 1936
Joseph J. Kadlec	1935 to 1936
Virgil Jorgenson	1935 to 1936
Chester Graupmann	1935 to 1936

Ben Turman	1935 to 1936
Theo. Ochu	1935 to 1936
John Schultz	1935 to 1936
Charles Arlt	1935 to 1943
Arthur Ohland	1935 to 1938
Ben Peik Secretary/Treasurer 1940 to 1946	1935 to 1946
C.A. Moore	1935 to 1935
Ed Boyle Vice President 1943 to 1946 Secretary/Treasurer 1946 to 1965	1935 to 1971
Peder Sondergaard	1935 to 1936
Ancher Nelson Vice President 1951 to 1953 Resigned to become REA Administrator	1936 to 1953
Carl Stender	1937 to 1938
Lynn Wulkan President 1948 to1960	1937 to 1961
Henry S. Hanson Vice President 1946 to 1948	1937 to 1948
Frank Haas	1937 to 1945
Ed Campbell	1938 to 1942
Jacob H.D. Hansen Vice President 1953 to 1963	1938 to 1963
Harry A. Bulau	1942 to 1952
Theodore Dietel	1943 to 1961
Arthur Sprengeler	1944 to 1952
Alvin Fischer	1945 to 1955
Melvin Todd	1946 to 1952
George Lhotka	1948 to 1952
Arvid Anderson	1948 to 1962
Victor Hahn	1950 to 1956
George Jungclaus	1951 to 1974
Victor Pulkrabek Vice President 1979 to 1980	1952 to 1982
Wayne Bulau Secretary/Treasurer 1965 to 1986	1952 to 1986
Walter Radke	1952 to 1960
Carl Ortloff President 1960 to 1970	1952 to 1970
Clarence Walter	1955 to 1968
G.E Birk	1953 to 1977

Name and Offices	Tenure
Orville Lipke President 1970 to 1983 Vice President 1963 to 1970, 1983 to 1985 CPA President 1968 to 1978	1956 to 1985
Leonard Winterfeldt	1960 to 1972
Albert Nicolai	1961 to 1973
Vernon Ernst Vice President 1970 to 1979	1961 to 1979
Donald Waldner Assistant Secretary/Treasurer 1971 to 1974	1962 to 1974
John Bernhagen	1963 to 1981
Vernon Ruschmeyer Assistant Secretary/Treasurer 1974 to 1983	1968 to 1983
Lester Ranzau Vice President 1980 to 1983, 2005 to 2009 President 1983 to 2005	1970 to date
Willard Strandquist	1971 to 1974
Burton Zimmerman	1972 to 1984
Duane Kulberg Vice President 1985 to 1988	1973 to 1988
Randall Thalmann	1974 to 1977
Gilbert Schwartz Assistant Secretary/Treasurer 1983 to 1989	1974 to 1989
Harold Toreen	1974 to 1977
Dale Dean	1977 to 1980
Marcel Mathison	1980 to 1989
Levi T. Tupa	1981 to 1987
Charles H. Jensen Vice President 1988 to 2001	1983 to 2001
Dale E. Peters Secretary/Treasurer 1987 to date	1984 to date
Charles Olesen Assistant Secretary/Treasurer 1989 to 1991	1985 to 1991
Edmund Ehrke	1987 to 2005
Roger J. Vacek	1987 to 1990
Curtis L. Rossow Assistant Secretary/Treasurer 1998 to 2000 Vice President 2003 to 2004	1988 to date
Mello L. Burns, Jr. Assistant Secretary/Treasurer 1991 to 1998	1989 to 1998
Virgil Stender	1989 to 1992

Roger Karstens Vice President 2001 to 2003, 2004 to 2005	1990 to date
Allan Duesterhoeft	1991 to date
Gerald Roepke Assistant Secretary/Treasurer 2000 to 2009 President 2009 to date	1992 to date
Kevin Maiers	1998 to 2001
Doug Kirtz President 2005 to 2009 Vice President 2009 to date	2001 to date
William Polchow Assistant Secretary/Treasurer 2009 to date	2001 to date
Oria Brinkmeier	2005 to date

Paul Theis, Legal Counsel Since 1989

Paul Theis has served as the legal counsel for McLeod Cooperative Power since 1989. Before that time he did some part-time work advising the Board of Directors and General Manager Harry Thiesfeld, when he was employed with the law firm of Hubert Smith. Hubert Smith was the attorney who served as McLeod Cooperative Power's legal counsel beginning in 1940.

Board working in 2009 meeting with NRECA representative.

CHAPTER 12
CURRENT EMPLOYEES AND BOARD OF DIRECTORS

In 2009, McLeod Cooperative Power employed thirty-two full-time and three part-time people. We are grateful for our dedicated staff that serves the needs of the MCPA membership.

The 2009 Board of Directors, seated left to right: Bill Polchow, Curtis Rossow, Dale Peters, Gerald Roepke, and Oria Brinkmeier; standing left to right: Allan Duesterhoeft, Doug Kirtz, Lester Ranzau, and Roger Karstens.

Back row, left to right, are linemen Brian Wika, Brad Hundt, Grant Miller, and Ben George; and line superintendent Mark Walford. Front row, left to right, are linemen Craig Marti, Curt Hanson, Dan Schade, Terry Underdahl, and Ryan Schuette.

Previous page: MCPA employees and family members participated in Glencoe's 2009 Holiday parade. A decorated bucket truck, employees as illuminated dancing trees and Christmas gifts, and our manager Kris Ingenthron Kringel were all part of the parade entry that tied for first place.

Administrative and Finance employees, left to right: Eric Sell, Pat Gavin, Carol Barlau, Teri Martin, general manager Kris Ingenthron, Jan Sanderson, and Randy Ahrndt.

Left: Customer Service and DIRECTV employees, left to right, are Becky Schiroo, Katie Ide, Jonathan Geiger, Patty Robb, Bob Thomes, Deb Goettl, Shannon Jerabek, Sue Pawelk, and Debbie Ebert.

Bottom left: Engineering Department employees, left to right, are Doug Kashmark, Laun Aiken, Greg Nistler, Robert Senst, Darrel Beste, and Dave Keil.

Not pictured: Part-time employees Sandy Tibbits, Janice Wark, and Julie Beneke.

CHAPTER 12 CURRENT EMPLOYEES AND BOARD DIRECTORS

ABOUT THE AUTHOR

Susan Pawelk

Susan "Sue" Pawelk has been the customer service manager at McLeod Cooperative Power Association since 1993. She is also the editor of the cooperative newsletter *McLeod Cooperative Power NEWS*. She enjoys working with McLeod's members, most of which she has found to be" pretty genuine and good, rural people."

Prior to MCPA, she was employed at Wright-Hennepin Electric Cooperative Association from 1984 to 1993. Before that she was a newspaper reporter and photographer for several years, working for the *Carver County News* in Watertown, Minnesota, and the *McLeod County Chronicle* in Glencoe.

Susan is a graduate of Mayer Lutheran High School and has a Bachelor's degree in Political Science from Hamline University in St. Paul, Minnesota. She spends her spare time in activities with her husband and sons. Heatwole Threshing Show is one of their favorite events. Susan enjoys photography as a hobby.

"I was honored to put together this history of McLeod Cooperative Power," said Pawelk. "Researching the great dedication of our Co-op's early founders and the great value early members placed on getting electricity, makes me appreciate the benefits from Co-op electricity so much more."